BIBLICAL MASSAGE AND HOLY SPIRIT TOUCH
With guidelines for
BIBLICAL MEDITATION AND FASTING

Biblical Massage and Holy Spirit Touch
With guidelines for
BIBLICAL MEDITATION AND FASTING

"Heal me O Lord, and I shall be healed; save me, and I shall be saved; for thou art my praise."
Jer. 17:14 KJV

He said, "If you listen carefully to the voice of the
LORD your God and do what is right in his eyes, if you pay attention to his commands and keep all his decrees, I will not bring on you any of the diseases I brought on the Egyptians, for I am the LORD, who heals you." Ex.15:26

Messenger K. Hezekiah Scipio
LMT, BCTMB, A.A.S, Massage Therapy, B.A. Psy,
Founder, Biblical Health Center Inc

Licensed Massage Therapist, MA59561
Board Certified in Therapeutic Massage and Bodywork

Email: delivergoodnews@gmail.com

Dedicated To
Ambassador Cyril B Boynes, jr
My sister-mother Wynta Ann Huggins
My brother, Duel Huggins
My sister, Bridget Williams
My sister Pastor Patricia Byrd-Konan
My brother, Pastor Nestor Konan
Mama Connie & Bishop Alexander Kissi
Thank you all.

delivergoodnews@gmail.com

Cover: Nancy Corniel, L.P.N., C.M.T

TESTIMONIAL

My name is Melody. I am an RN. I work with orthopedic, neuro, trauma patients, and a variety of other patients. I have seen those hurting with physical wounds; I have also seen those with emotional, mental and spiritual pain. The wounds affecting their mental and emotional conditions can sometimes cripple them in ways more severe than the physical wounds. I was curious and intrigued to get Messenger Hezekiah's book on healing touch. I read the book through and found it interesting. I know that healing can occur, and scripture tells us to lay hands on the sick. I long to see more healing by the power of our Lord and Saviour Jesus Christ Yeshua through the gift of the Holy Spirit promised all believers. Messenger Hezekiah gives scriptural basis for this healing and talks about the faith and requirements laid out in scripture for such a work. I believe in this purpose and the need for those trained in this area. May God bless those who are called to do a work of this sort for His glory.
~Melody
February 9, 2019

Table of Contents

*****

Is anyone among you sick?
"Is anyone among you suffering? Let him pray. Is
anyone cheerful? Let him sing praise. Is anyone
among you sick? Let him call for the elders of the
church, and let them pray over him, anointing him
with oil in the name of the Lord. And the prayer of
faith will save the one who is sick, and the Lord will raise him up. And
if he has committed sins, he will be forgiven. Therefore, confess your
sins to one another and pray for one another, that you may be healed.
The prayer of a righteous person has great power as it is working."
(James 5 13-16)

*****

Introduction

I wrote and self-published my first book in 2007. It was titled, *God's Amazing Bible Plants Healed Me.* I wrote it to testify for God's faithfulness in keeping his promise as reported in Ezekiel 47:12. I wrote it on my recovery from illnesses of unknown origin that had my doctors baffled, and conclude I was not going to survive. I published my second book , *Biblical Massage and Holy Spirit Touch* in March 2010 out of my frustration that curricula of Massage Therapy Institutions were mainly techniques from Eastern religious principles; so, I was determined to demonstrate that Massage Therapy , or what I call Biblical Massage , has historical and biblical connections with Judeo-Christian (w)holistic healing environment. In this revised edition, *Biblical Massage and Holy Spirit Touch with guideline for Biblical Meditation and Fasting*, my emphasis is on Biblical Meditation and Fasting. I define Biblical Meditation, explain differences Biblical Meditation and Transcendental Meditation, and detail step by step methods of performing Biblical Meditation.

I am an ordained minister of God in Christ Jesus. I obtained my Bachelor of Arts degree in Psychology, from Argosy University, Tampa, Florida, Associate of Applied Science degree in Massage Therapy from Siena Heights University in Adrian, Michigan, and licensure as Massage Therapist from Sanford Brown Institute, Tampa, Florida. I had my Board Certification in Therapeutic Massage and Bodywork in 2010 from the National Certification Board for Therapeutic Massage and Bodywork. My status remained active until May 2017 when I allowed renewal of my license to lapse to enable me cope with the stress of undergraduate studies in psychology at Argosy University, Tampa, Florida.

Studying massage therapy was in response to a personal conviction that I was called by Christ Jesus to go, preach, teach the gospel, and lay hands upon sick people, and they will be healed. (Mk 16:18) But during my second module, I became so overwhelmed by doubts that I questioned if it was biblical at all for Christians to give or receive massage, considering the invasiveness of touch therapy. I was troubled even more that Asian Bodywork Therapy, including Ayurveda and

Energy-Based Bodywork Therapy, as well as Reiki have world acclaimed acceptance and recognition mainly on account that "they have been in practice for thousands of years." Significant parallel credits have also been lavished upon the religious traditions and principles that produced and govern the modalities which include yoga and tai chi; but there is a dearth of such comparable citation for the real Author and Healer of our Spirit, soul, and body, Almighty God, Who gave Christ Jesus authority to heal the sick by the Touch of the Holy Spirit. I had felt discouraged and, on many occasions, I was very close to withdrawing from class, but thanks to a perceptive instructor, Ms Channie Bell, (who provided me an opportunity of insight into Qigong), I graduated, was licensed, and Board Certified in Therapeutic Massage and Bodywork.

I am thankful that God, in His own appointed time, has given me the grace of insight into what constitutes Biblical Massage and Holy Spirit Touch which he has also inspired his prophets, scribes, and apostles to record in both the Old and New Testament scriptures. In time, I came to realize that some of the doubts about massage therapy arose from my ignorance of evidence-based health benefits of palpating and manipulating soft tissues, and, in part, also, from my unease at the inroads of Eastern religious philosophies into many areas of massage therapy to the seclusion of Judeo-Christian beliefs. But now, I have learned to identify and separate biblically supported techniques or modalities from what the scriptures forbid, while noting historical facts that allopathy was, not so long ago, a discredited practice and the term "allopathy" was applied pejoratively to the extent that people practicing medicine were burnt to death for supposedly practicing witchcraft although ancient Greeks deified Asclepius, "father of medicine and healing;"(Encyclopedia Britannica) his rod, a snake-entwined staff remains a symbol of medicine today, sometimes without the snake. In all of these, I have learned sufficiently to recognize the baby in the bathtub and not to throw both away.

Yesterday, I kept an appointment with my urologist, and was diagnosed with prostate cancer. My response to this piece of diagnostic information is shaped by Paul's letter to the Philippians, chapter 1:21, "For to me to live is Christ, and to die is gain."

I am not concerned about what the prognosis will be. If God wants me to continue living, and if living will do some good to people for His glory, God, who is The Healer of all diseases, will lead me to recovery through Christ Jesus. But if He wants me home with Him, so be it. What matters to me is to live a life that does not dishonor the name of Jesus Christ, so that when I die, my death will be profitable. So, brothers and sisters, as Biblical Massage Ministers/Therapists/Practitioners, you and I are called to represent Christ. It is for this reason God will use us as His instruments to heal others. He demands from us to focus exclusively on Him. We must proclaim and acknowledge Him as the One who heals. He will not allow His Spirit that is indwelling in us to be shared or mixed up in the same vessels, our human bodies, in any way with any other spirit. God says, "I am Yahweh, that is my name; my glory I give to no other, nor my praise to idols." (Isaiah 42:8) Our God is One. We must study to meet all requirements of our educational authorities in order to pass every necessary examination by the laws of the State and the country. For the scriptures demand from us obedience to the laws of the land. When we have passed our examinations and are in full possession of our licenses, it is incumbent upon us to obey God and take a stand for Jesus. For God will judge us by the choices we make through life.

"And I saw the dead, great and small alike, standing before the throne. Books were opened, and then another book was
opened, the book of the living. The dead were judged according to what they had done, as recorded in the books. Then the sea gave up its dead. Death and the world of the dead also gave up the dead they held. And all were judged according to what they had done.
Then death and the world of the dead were thrown
into the lake of fire. (This lake of fire is the second
death.) Whoever did not have his name written in the book of the living was thrown into the lake of fire." (Revelation 20:12-15)

The Holy Spirit will come and dwell in those who believe in Jesus Christ if they ask Him for it. Whenever we are called upon to attend to the dying, it is our responsibility to reassure the dying of a place in eternity where there will be no more sorrow, no more weeping, no more pain; a place of eternal joy that God has reserved for all of us who believe in His Son Jesus Christ. The unbeliever who accepts Jesus Christ as his personal Savior, even though he does so in his dying moment, can be sure of his or her salvation. The bible says when Jesus was crucified, two criminals were crucified also, one on Jesus' right side and the other on His left side. One of the two, mocked and insulted Jesus. The other, rebuked the irreverent criminal, "And he said unto Jesus, Lord, remember me when thou comest into thy kingdom. And Jesus said unto him, Verily I say unto thee, today, shalt thou be with me in paradise." (Luke 23:42-43)

Whenever appropriate, proclaim this message of hope; lead the patient to faith in Jesus Christ as the only way to God the Father. Say the prayer of salvation to heal the sick in the Spirit, mind and body, to take away the fear of dying; and help provide that person the knowledge that in death, God has promised resurrection for everyone who believes to be in union with Christ Jesus. Help restore the spirit of calmness and peace to the dying so that, like Apostle Paul, they too can confidently say, "I have fought the good fight, I have finished the race, I have kept the faith. From now on, there is reserved for me the crown of righteousness, which the LORD, the Righteous Judge, will give me on that day, and not only to me but also to all who have longed for His appearing." Amen! (2 Timothy 4:7)

*****

I will restore health to you
"For I will restore health to you, and your wounds I will heal, declares the Lord, because they have called you an outcast: 'It is Zion, for whom no one cares!'" (Jeremiah 30:17)

*****

2
Biblical Massage and Holy Spirit Touch

Biblical Massage is a healing technique that trusts Almighty God as the Sovereign Healer of the sick through Jesus Christ by the Power of the Holy Spirit. Research suggests "religious belief in a concerned GOD can improve response to medical treatment for depression," a study at Rush University Medical Center, Chicago, concluded, according to a paper in the Journal of Clinical Psychology published February 24, 2010.

When the Lord Jesus commanded his followers to preach the gospel and lay hands on the sick and they would recover, according to Mark's gospel chapter 16:18, his followers since then maintained a tradition known as 'laying on of hands," which is also known to massage therapy professionals as "passive touch." The "laying on of hands" or "passive touch" was just one component in a blend of techniques that (w)holistic healers practiced for thousands of years to restore health to the sick in the spirit, mind, and body. It also involved palpating and manipulating soft tissue by compressing, squeezing, and pushing the body muscles of the patient over prayers enabling blood to circulate from the "crown of the head to the sole of the feet." The practitioner also used oils, perfumes, music, and prayers, while the patient, having fasted, meditated on things noble and pure and of good report, in the belief that ultimately, Almighty God was the One who healed. James, the brother of the Lord Jesus, summarized this method of healing in his letter to believers all over the world as follows: "Is anyone among you in trouble? He should pray. Is anyone happy? He should sing songs of praises. Is there anyone who is sick? He should send for church elders, who will pray for him and rub olive oil on him (rub is act of massaging, not pouring oil) in the name of the Lord. This prayer made in faith will heal the sick person; the Lord will restore him to health, and the sins he has committed will be forgiven" (James 5:13-16). In our world and time, practitioners of this "healing art" are called "Massage Therapists." Nearly 400 years before the Lord Jesus, a young Jewish woman called Hadassah or Esther, a captive in the harem of Iranian/Persian king Xerxes, submitted herself to intensive relaxation massage treatment for one whole year in preparation for a nationwide beauty pageant which won her the trophy of a crown to become Queen of Iran/Persia. The Good News Bible Today's English Version states this: "He lost no time in beginning her treatment of massage and special diet... The regular beauty treatment for the women lasted a year -- massages with oil of myrrh for six months, and with oil of balsam for six months"(Esther 2: 8 -12).

About 400 years later, the Lord Jesus was the center a certain massage therapist's love and attention ; the woman gave the Lord Jesus a scalp massage, reflexology and aromatherapy, part of massage therapy techniques. Some theologians claim the woman was Mary Magdalene. King James Version of Luke 7:36-38 does not adequately portray the Lord Jesus' posture to show how the woman while standing behind the Lord Jesus at table eating could weep tears on Jesus' feet and wipe his feet with her hair (Luke 7:37-38). No matter the length of the woman's hair, the action King James version describes is impossible, until one reads the versions from Greek or Aramaic, the language the Lord Jesus spoke and which indicates that the Lord Jesus "reclined," a sitting posture which suggested the Lord Jesus resting on couch-like piece of furniture, his legs outstretched; massage therapists can recognize this posture, and know it is well suited for the body mechanics of the woman   described   in   the   text.

This book cites scientific theories and examples to back up the reason 'Biblical Massage and Holy Spirit Touch with Biblical Meditation and Fasting' may be considered effective treatment for emotional and muscular pain. It includes step by step methods of combining Biblical Massage with Biblical Meditation and Fasting.

Biblical Massage and Holy Spirit Touch focuses on healing the spirit of man, his soul, and body, using biblical meditation, Christian hymns of praises and worship, fasting, prayers, scriptures, oils, diet, nutrition, herbal remedies, and soft tissue manipulation, including: Deep Tissue, Myofascial Release, Lymphatic Drainage, Proprioceptive Neuromuscular Facilitation (PNF), Muscular Energy Technique (MET), and Trigger Point Therapy while integrating techniques of Western Massage and isometric exercises.

Biblical Massage has been practiced in the Middle East and the Mediterranean geographical areas and among Jewish people for ages although no one called the treatment Biblical Massage. I plead guilty for the name. Call this field of (w)holistic healing what you may, it baffles the mind that despite its longevity, for whatever reason, text books have ignored giving it the recognition and credit due its Creator, Almighty God. It is interesting to note that application of massage therapy as remedy for pain, was acknowledged following a 2018 study by team of researchers from the National Institutes of Health, National Center for Complementary Health. The team of researchers including Perlman A, Fogerite SG, Glass O, et al, (2018), concluded, "a weekly session of massage therapy may provide short-term benefits for people with osteoarthritis of the knee, including reduced pain and stiffness and improved function," according to a new NCCIH-funded study. The study, which was done at locations in North Carolina, New Jersey, and Connecticut, was published in the Journal of General Internal Medicine.

Researchers randomly assigned 222 people with knee osteoarthritis to receive whole-body Swedish massage, light touch, or usual care. Light touch involved the massage therapist gently placing his or her hands in a specified sequence on the participant's major muscle groups and joints. Usual care was the participant's typical care regimen for osteoarthritis. Participants in the massage or light touch groups received 8 weekly treatments (lasting 60 minutes each), and then were randomly assigned to receive treatment every 2 weeks or usual care to week 52.

The researchers found that 8 weeks of massage provided statistically and clinically significant improvement of osteoarthritis symptoms, as assessed by a widely used questionnaire that evaluates the condition of people with arthritis. Pain, stiffness, and physical function all improved. Side effects were minimal.

The researchers noted that these findings support the results of their previous research in people with knee osteoarthritis, which demonstrated the safety, feasibility, and potential efficacy of 8 weeks of massage therapy in improving pain, stiffness, timed 50-foot walk, and physical function compared to a waitlist control. However, long-term benefits were less clear; after 8 weeks of treatment, massage therapy-maintained improvement but did not provide additional benefit beyond usual care. The researchers noted that the study size was relatively small, and the participants were mostly white women. The results may not be generalizable to other groups of people. (Journal of General Internal Medicine, December 12, 2018).

Biblical Massage and Holy Spirit Touch is (w)holistic healing, coalescing the Spirit, mind and body together. In his paper entitled Orthodox Theological Roots of Holistic Healing, John Chirban, PhD, ThD, Director of the Institute of Medicine, Psychology, and Religion in Cambridge, Massachusetts, at Harvard Medical School and Theology, Cambridge Hospital, writes, "Holistic healing, which was part of the established Jewish tradition, had its earliest roots in the miracle accounts of healings by Jesus Christ. The Byzantine people placed healing in a scriptural context, understanding of which suffused Byzantine culture through icons, pictures, and all forms of art. And, as in the Bible, healing could have a miraculous quality: "Everything is possible to those who believe" (Mt 8:13; 21:22; Mk 9:23; Lu 8:50) Dr. Chirban goes on, "Nothing is more certain than the fact that Jesus was a healer. No information about Jesus of Nazareth is so widely and repeatedly attested in the New Testament gospels as the fact that he was healer of people who suffered from physical, mental, and spiritual distress. From the synoptic tradition we hear that, "They brought to him all who were ill or possessed of by devils; and the whole town was there, gathered at the door. He healed many who suffered from various diseases and drove out many devils..."

All through Galilee he went, preaching in the synagogues and casting out the devils...He cured so many that sick people of all kinds came crowding in upon him to touch him. Jesus can be seen as a physician. Indeed, three times he is reported to have used the term physician self-referentially. Mark reports that when he was criticized for dining with tax collectors and sinners he responded, "It is not the healthy who need a physician but the sick."

The Lord Jesus Himself might have received a massage therapy during a stressful time in his life. Luke's description in Luke 7:36-50 of an event in the life of the 100% God 100% human, the Lord Jesus, invokes

16

the picture of a woman treating the Lord Jesus to aromatherapy combined with reflexology and scalp massage in act of love.

King James' version of Luke's account does disservice to Western mindset by depicting the activity of Jesus in the home of the dignitary, Shimeon as "sat at meat with him," which to Western minds seem like sitting at a table in a restaurant, or , at dinner table in Western homes. Here is King James version of the text : "36 And one of the Pharisees desired him that he would eat with him. And he went into the Pharisee's house, and sat down to meat. Verse 37. And, behold, a woman in the city, which was a sinner, when she knew that Jesus sat at meat in the Pharisee's house, brought an alabaster box of ointment,38 And stood at his feet behind him weeping, and began to wash his feet with tears, and did wipe them with the hairs of her head, and kissed his feet, and anointed them with the ointment.

39 Now when the Pharisee which had bidden him saw it, he spake within himself, saying, This man, if he were a prophet, would have known who and what manner of woman this is that toucheth him: for she is a sinner.40 And Jesus answering said unto him, Simon, I have somewhat to say unto thee. And he saith, Master, say on.41 There was a certain creditor which had two debtors: the one owed five hundred pence, and the other fifty. 42 And when they had nothing to pay, he frankly forgave them both. Tell me therefore, which of them will love him most? 43 Simon answered and said, I suppose that he, to whom he forgave most. And he said unto him, Thou hast rightly judged. 44 And he turned to the woman, and said unto Simon, Seest thou this woman? I entered into thine house, thou gavest me no water for my feet: but she hath washed my feet with tears, and wiped them with the hairs of her head. 45 Thou gavest me no kiss: but this woman since the time I came in hath not ceased to kiss my feet.46 My head with oil thou didst not anoint: but this woman hath anointed my feet with ointment.47 Wherefore I say unto thee, Her sins, which are many, are forgiven; for she loved much: but to whom little is forgiven, the same loveth little. 48 And he said unto her, Thy sins are forgiven. 49 And they that sat at meat with him began to say within themselves, Who is this that forgiveth sins also? 50 And he said to the woman, Thy faith hath saved thee; go in peace".

However, both the New International Version and Aramaic Version in the native language of the Lord Jesus back the Greek-English interlinear translation in using the word "reclined" to invoke a mental picture of how the Lord Jesus "sat at meat," which was one of three customary positions the male of Middle East Mediterranean culture "sat at meat;" they might sit lotus like position on a mat, animal skin or rug spread on the floor, sit on a chair to dine at table, or, lounge on couches with their legs stretched out as grapes, dates , or fruits of choice were passed around while they discussed issues that mattered; "reclined" conveys the picture. It is the word of choice in the NIV: "36 One of the Pharisees asked him to eat with him, and he went into the Pharisee's house and reclined at table. 37 And behold, a woman of the city, who was a sinner, when she learned that he was reclining at table in the Pharisee's house, brought an alabaster flask of ointment, 38 and standing behind him at his feet, weeping, she began to wet his feet with her tears and wiped them with the hair of her head and kissed his feet and anointed them with the ointment. 39 Now when the Pharisee who had invited him saw this, he said to himself, "If this man were a prophet, he would have known who and what sort of woman this is who is touching him, for she is a sinner." 40 And Jesus answering said to him, "Simon, I have something to say to you." And he answered, "Say it, Teacher."

41 "A certain moneylender had two debtors. One owed five hundred denarii, and the other fifty. 42 When they could not pay, he cancelled the debt of both. Now which of them will love him more?" 43 Simon answered, "The one, I suppose, for whom he cancelled the larger debt." And he said to him, "You have judged rightly." 44 Then turning toward the woman he said to Simon, "Do you see this woman? I entered your house; you gave me no water for my feet, but she has wet my feet with her tears and wiped them with her hair. 45 You gave me no kiss, but from the time I came in she has not ceased to kiss my feet. 46 You did not anoint my head with oil, but she has anointed my feet with ointment. 47 Therefore I tell you, her sins, which are many, are forgiven—for she loved much. But he who is forgiven little, loves little." 48 And he said to her, "Your sins are forgiven." 49 Then those who were at table with him began to say among themselves, "Who is this, who even forgives sins?" 50 And he said to the woman, your faith has saved you; go in peace."

Here is also the Aramaic version in English, and note the words in bold, underlined for emphasis. "36 But one of the Pharisees came asking him to eat with him and he entered the Pharisee's house and he reclined. 37 And a sinner woman who was in the city, when she knew that he was staying in the Pharisee's house, she took an alabaster vase of ointment. 38 And she stood behind him at his feet, and she was weeping, and she began washing his feet with her tears and wiping them with the hair of her head. And she was kissing his feet and anointing them with ointment. 39 But when that Pharisee who had invited him saw, he thought within himself and he said, "If this one were a Prophet, he would have known who she is and what her reputation is, for she is a sinner woman who touched him." 40 But Yeshua answered and he said to him, "Shimeon, I have something to tell you." But he said to him, "Say it, Rabbi." And Yeshua said to him, "One landowner had two debtors; one debtor owed him 500 denarii and the other 50 denarii." "And because they had nothing to pay he forgave both of them. Which of them therefore will love him more?" 43 Shimeon answered and he said, "I suppose he who was forgiven most." Yeshua said to him, "You have judged correctly." 44 And he turned to that woman and he said to Shimeon, "Do you see this woman? I entered your house, yet you gave no water for my feet and she has washed my feet with her tears and has wiped them with her hair." 45 "You did not kiss me, but look, from when she entered, she has not ceased to kiss my feet." 46 "You did not anoint my head with oil, but this one has anointed my feet with oil of ointment." 47 "On account of this, I say to you, that her many sins are forgiven her because she loved much, but he who is forgiven a little loves a little." 48 And he said to her, "Woman, your sins are forgiven you." 49 But they who were reclining said in their souls, "Who is this that he even forgives sins?" 50 But Yeshua said to that woman, "Your faith has given you life, go in peace."

Consider this Greek-English interlinear, and note verse 36 line two keywords, "he reclined," and verse 37 line two keywords, "he had reclined.

In the King James version, the words, and sat down to meat invoke the picture of Michael Angelo's image of 'the Last Supper,' or, a feature in a Western restaurant or home, however, and he reclined seems an accurate depiction reclining on a piece of furniture imitating a massage table in appearance as can be deduced from the following pictures. Bear in mind that Pharisees in the era of the Lord Jesus, were held in high esteem; they might not sit at the table to dine as Westerners do, but on a couch with feet outstretched – reclined. Furniture might be arranged in a room in a way to allow space between the wall and couches so that the woman, who, some theologians say, was Martha's sister, had the convenient space to move around and could practicably wash the feet of the Lord Jesus with her tears, dry them with her hair , and perform aromatherapy on his head. The body mechanics is more plausible with Jesus reclining on a couch than sitting at table. Thus, it is safe to safe that the Lord Jesus Himself received massage therapy.

The pictures of typical couches in wealthy priests' home in Jerusalem I have posted below look very much like massage tables, and the type one might use while standing to perform foot and scalp massages. The woman who washed the feet of the Lord Jesus with her tears and wiped it with her hair, was likely a traveling massage therapist who was able to save up money to afford the high priced perfume bottled in a precious alabaster-clay container which she had kept specially for the Lord Jesus. As was the view of the time even as it is today in many countries, a woman who was a traveling massage therapist was considered a 'sinner;' in other words, she was thought to be a prostitute, and Shimon, the host thought Jesus was supposed to know the woman's ill-reputation; instead, the Lord Jesus praised her.

RECONSTRUCTION of weathy ancient Roman Priest's home in Jeruzalem by Gregori Louis

Reconstruction of a Wealthy Priest's Home in Jerusalem's Upper City

Structure on which to rest or sleep. The Hebrew term מטה, meaning "divan" as well as "bed," is synonymous with ערש(Amos iii. 12) and משכב(II Sam. xvii. 28). In olden times the Jewish bed, a plain wooden frame with feet, and a slightly raised end for the head (Gen. xlvii. 31), probably differed little from the simple Egyptian bed. The frame covered with מרבדים(Prov. vii. 16), served as a bed for the old and sick during the day (Gen. xlvii. 31; I Sam. xix. 13 et seq.), while at meals people sat on it, perhaps with crossed legs (compare Ezek. xxiii. 41; I Sam. xx. 25).Source: Emil G. Hirsch & Wilhelm Nowack

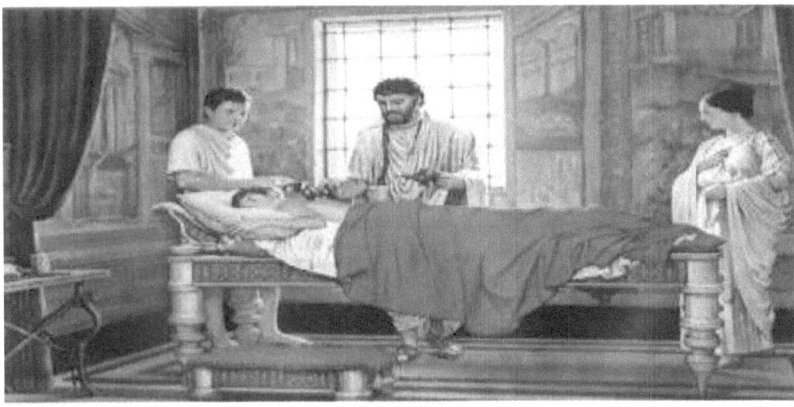

Artist's reconstruction of Roman healing ca 140-120BC -Source: Pinterest

Luke		7		Version		Commentary		Language	

A Sinful Woman Anoints Jesus

Ἠρώτα δέ τις αὐτὸν τῶν Φαρισαίων, ἵνα φάγῃ μετ᾽ αὐτοῦ· καὶ

εἰσελθὼν εἰς τὸν οἶκον τοῦ Φαρισαίου, κατεκλίθη. καὶ ἰδοὺ, γυνὴ ἥτις

ἦν ἐν τῇ πόλει, ἁμαρτωλός, καὶ ἐπιγνοῦσα ὅτι κατάκειται ἐν τῇ

οἰκίᾳ τοῦ Φαρισαίου, κομίσασα ἀλάβαστρον μύρον, καὶ στᾶσα ὀπίσω παρὰ

Recent archaeological discoveries present proofs of contrast hydrotherapy, such as the use of hot and cold springs in ancient cities of modern Turkey, (Asia Minor of the Bible) which included three cities cited in the Bible, and they were, Pergamon, Hierapolis, and Laodicea. In Revelation 2: 12-13 our Lord Jesus Christ called Pergamon "Satan's Throne," Pergamon was noted for its medicine. Like many Greek cities, Pergamon was polytheistic, erecting massive temples for the city's gods, Zeus, Athena and others. Pergamon also boasted a Sanctuary dedicated to Asclepius, "god of healing." Sarah K. Yeomans , Managing Web Editor of Biblical Archaeologist Review and the Travel Study Director of the Biblical Archeology Society writes, "In this place people with health problems could bathe in the water of the sacred spring, and in the patients' dreams Asclepius would appear in a vision to tell them how to cure their illness." (Yeomans, Biblical Archeologist Review)

Pergamon rose to prominence during the years of the Greek empire's division following the death of Alexander the Great in 323 B.C. Hierapolis was the name of another neighboring city with a thermal spa. The city "became a healing center where doctors used the hot thermal springs as a treatment for their patients. People came to soothe their ailments, with many of them retiring or dying here...." (Yeomans) The third was Laodicea, a very wealthy city in ancient Turkey (Asia Minor). Founded between 261 and 253 B.C. by Antiochus 11, Theos of Syria. Laodicea became a booming center for banking and finance, manufacturing of clothing and dyeing. It had an elite school of medicine where a famous ophthalmologist practiced and was noted for its production of ingredients used in the concoction of eye and ear medications. "One of the principles of medicine at that time was that compound diseases required compound medicines. One of the compounds used for strengthening the ears was made from the spice nard (an aromatic plant). Galen says that it was originally made only in Laodicea, although by the second century A.D. it was made in other places also. Galen also described a medicine for the eyes made of Phrygian stone...." (Blake and Edmonds, Biblical Sites in Turkey, p. 140) No other city...was as dependent on external water supplies as Laodicea. Water was also piped in through an aqueduct from Colossi...The luke-warmness for which... the name of Laodicea has become proverbial, may reflect the condition of the city's water supply. The water supplied by the spring... was tepid and nauseous by the time it was piped to Laodicea, unlike the therapeutic hot water of Hierapolis or the refreshing cold water of Colossae (The Anchor Bible Dictionary Rudwick and Green 1958). Therefore when in Revelation 3:15-16 , our Lord Jesus Christ in His message through John to "the angel of the church in Laodicea" said, " I know that you are neither cold nor hot; how I wish you were either one or the other; I am going to vomit you out of my mouth," He was speaking in metaphors that were true of Laodicea familiar to John's intended audiences. Water piped into Laodicea by aqueduct from the south was so concentrated with minerals that it became lukewarm and made the air around Laodicea nauseous. "Roman engineers designed vents, capped by removable stones, so the aqueduct pipes could periodically be cleared of deposits." (John McRay, Archaeology And The New Testament, p. 248).

Prophet Elisha, before the time of Jesus, ordered a leprous Syrian General, Naaman by name, to undergo hydrotherapy in the Jordan River. 2 King 5:14 states, "...Then went he down, and dipped himself seven times in the Jordan, according to the saying of the man of God; and came again like unto the flesh of a little child, and he was clean." Blueletterbible dot org states as follows: "Pool of Siloam is a rock-cut pool on the southern slope of the City of David (believed to be the original site of Jerusalem) now outside the walls of the Old City to the southeast. The pool was fed by the waters of the Gihon Spring, which were carried there by two aqueducts - a 20ft deep direct cutting that was covered with rock slabs, and dates from the Middle Bronze Age ~1800BC, and Hezekiah's Tunnel, a curving tunnel within the bedrock, dating from the reign of King Hezekiah~700BC.

The prophet Isaiah refers to the pool's waters in Isaiah 22: 10 - 11; "You inspected all the houses in Jerusalem and tore some of them down to get stones to repair the city walls. In order to store water, you built a reservoir inside the city to hold the water flowing down from the old pool. But you paid no attention to God, who planned all this long ago and who caused it to happen." (Isaiah 22:10-11) For Christians, the pool has additional significance as it is mentioned in the Gospel of John, as the location to which Jesus sent a man who had been blind from birth... The place was called Bethesda or "house of mercy." It contained a pool which was located near the Sheep Gate. The pool area was covered with five porticoes. The porticoes were covered areas where the sick would lie and wait. "In the 4th century, and probably down to the Crusades, a pool was pointed out as the true site, a little to the Northwest of the present Stephen's Gate; it was part of a twin pool and over it were erected at two successive periods two Christian churches. Later on, this site was entirely lost and from the 13th century the great Birket Israel, just north of the Temple area, was pointed out as the site. Within the last quarter of a century, however, the older traditional site, now close to the Church of Anne, has been rediscovered, excavated and popularly accepted. This pool is a rock-cut, rain-filled cistern, fifty-five feet long by twelve wide, and it is approached by a steep and winding flight of steps. The site still attracts the sick. They go expecting healing." (Bible Atlas, Encyclopedia,

Bethesda, Jerusalem)

During the time of our Lord Jesus Christ, the bible says people went down to the Pool of Siloam and waited in line to be healed of their sicknesses after an angel came into it and "troubled the waters." The first person to get into the pool was healed of his or her illness. The belief that any sick person who was the first to enter the water after an angel had come to "trouble the water" got healed, goes to support the fact that the Jewish people even long before the era of Christ Jesus, believed that the sick was healed through Divine intervention, and not by any innate healing Force; if that was not hydrotherapy, tell me, I want to know. It is noteworthy also that Christian churches or monuments were built over those locations in later years. Contemporary Western Massage may find similarity in the treatment Esther received before participating in the beauty pageant for the prize of the crown of Persia. In 407BC, a Jewish captive, Hadassah or Esther, was a war captive in the harem of the Persian ruler, Ahasuerus. It is stated in Esther 2:9 &12 TEV that, "Hegai liked Esther, and she won his favor." He lost no time in helping Esther prepare for a beauty pageant beginning with treatment of massage and special diet. "The regular beauty treatment for the women lasted a year –massages with oil of myrrh for six month and with oil of balsam for six more."

Oils of different types, most especially olive oil, were used for healing and for massage. Oil has been used as a facial ointment; the Bible states, "You cause the grass to grow for the cattle, and plants for people to use for food from the earth, and wine to gladden the human heart, oil to make the face shine..." (Ps 104:14-15)

*****

Fear not for I am with you!
"Fear not, for I am with you; be not dismayed, for I am your God; I will strengthen you, I will help you, I will uphold you with my righteous right hand."(Isaiah 41:10)

*****

3
Laying on of Hands

The scriptures give numerous examples when laying on of hands was used to heal the sick. The Lord Jesus Christ Himself commanded his followers, as it is reported in Mark 16:15-18, "And he said unto them, "Go ye into all the world, and preach the gospel to every creature. He that believeth and is baptized shall be saved, but he that believeth not, shall be damned. And these signs shall follow them that believe. In my name they shall cast out devils; they shall speak with new tongues. They shall take up serpents, and if they drink any deadly thing, it shall not hurt them; they shall lay hands on the sick, and they shall recover."

Raising The Dead
Elijah
1 Kings Chapter 17 from verse 1 to 23. This is what the scripture says: (Elijah) "cried out to the Lord, 'O Lord my God, have you brought calamity even upon the widow with whom I am staying, by killing her son?" Then he stretched himself upon the child three times, and cried out to the Lord, "O Lord my God let this child's life come into him again." The Lord listened to the voice of Elijah; the life of the child came into him again, and he revived."

Elisha
2 Kings 4:32-35, "When Elisha arrived, he went alone into the room and saw the boy lying dead on the bed. He closed the door and prayed to the Lord. Then he lay down on the boy, placing is mouth upon his mouth, and his eyes upon his eyes, and his hands upon his hands, and stretched himself upon the child. As he lay stretched out over the boy, the boy's body started to get warm. Elisha got up, walked around the room, and
then went back and again stretched himself over the boy. The boy sneezed seven times and then opened his eyes."

Peter
Acts 9:36-40: "In Joppa there was a woman named Tabitha, who was a believer... At that time, she got sick and died...Joppa was not very far from Lydia, and when the believers in Joppa heard that Peter was in Lydda they sent two men to him... When he arrived, he was taken to the room upstairs... Peter knelt down and prayed;

then he turned to the body and said, Tabitha, get up!" She opened her eyes, and when she saw Peter, she sat up..."

Paul

Acts 20: 7 –10: "On Saturday evening, we gathered together for the fellowship meal. Paul spoke to the people and kept on speaking till midnight... A young man named Eutychus was sitting in the window, and as Paul kept on talking,

Eutychus got sleepier and sleepier, until he finally

went sound asleep and fell from the third story to the ground. When they picked him up, he was dead. But Paul went down and threw himself on him and hugged him. "Don't worry, he said, he is still alive."

*****

Gifts of healing by that one Spirit "to another faith by the same Spirit, to another gifts of healing by that one Spirit." (1 Corinthians 12:9)

*****

4
Healing, A Spiritual Gift

Who Can Perform Biblical Massage?

The Minister/Therapist must be a Christian. Why?

This is why. The Bible asks, "How can light and darkness live together? How can Christ and the Devil agree? How can God's temple come to terms with pagan idols? For we are the temple of the living God" (2 Corinthians 6: 14 – 16).

When an unbeliever converts to Christianity, God regenerates him by giving him or her a spiritual new birth. The Holy Spirit then comes and dwells in him. The Bible teaches that healing, that is, the supernatural ability to bring or release health to a person's spirit, soul, and body is spiritual gift. The Holy Spirit gives spiritual gifts liberally for service to the glory of God.

In 1 Corinthians 12:4-11, it is stated, "Now, there are varieties of gifts, but the same Spirit and there are varieties of services, but the same Lord, and there are varieties of activities, but it is the same God who activates all of them in everyone. To each is given the manifestation of the Spirit for the common good. To one is given through the Spirit the utterance of wisdom, and to another the utterance of knowledge according to the same Spirit. To another faith by the same Spirit, to another gifts of healing by the same Spirit, to another, the working of miracles, to another prophesy, to another the discernment of spirits, to another various kinds of tongues, to another the interpretation of tongues. All these are activated by the one and the same Spirit, who allots to each one individually just as the Spirit chooses."

The spiritual gifts, as defined in 1 Corinthians 12, by an unknown theologian, "enable a witness to others as to the validity of Christ by means of supernatural workings in ways that could only be done by Him. Jesus Christ commands us to continue witnessing for Him; spiritual gifts are God given means to witness for Jesus. This is one thing we see with Christ as he ministered here on earth (Acts 2:22). These too were signs to both believers and unbelievers. So, following that same thought pattern, we too have gifts to testify, glorify, and validate Christ as authentic and personal. "Spiritual gifts are for today in order to carry out the mission of the church which is seen in Matthew 28:19 and Acts 1:8." (John Piper)

*****

He himself bore our sins in his body on the tree
He himself bore our sins in his body on the tree, that we might die to sin and live to righteousness. By his wounds you have been healed. (1 Peter 2:24)

*****

5
Finding Your Spiritual Gifts

John Piper, pastor of Bethlehem Baptist Church in Minneapolis, Minnesota, writes, "The apostle Paul wrote to the Corinthians, 'Now concerning spiritual gifts, brothers, I do not want you to be uninformed" (1 Corinthians 12:1) ... Instead of spreading myself too thin across 1 Corinthians 12,13 and 14 (the major section on spiritual gifts) I have chosen to focus on several smaller texts so that we can examine their teaching more closely. If you were reading through the New Testament, the first place you would run into the term "spiritual gift" is Romans 1:11, 12. ... Writing to the church at Rome, Paul says, "I long to see you, that I may impart to you some spiritual gift to strengthen you, that is, that we may be mutually encouraged by each other's faith, both yours and mine." The translation "impart to you some spiritual gift" is misleading because it sounds like Paul wants to help them have a gift, but the text actually, means that he wants to give them the benefit of his gifts. "I long to see you that I may use my gifts to strengthen you." The first and most obvious thing we learn from this text is that spiritual gifts are for strengthening others. This, of course, does not mean that the person who has a spiritual gift gets no joy or benefit from it. ... But it does suggest that gifts are given to be given. They are not given to be hoarded. "I desire to share with you some spiritual gift to strengthen you." What does strengthen mean? He's not referring to bodily strength but strength of faith. The same word is used in 1 Thessalonians 3:2 where Paul says, "We sent Timothy, our brother and servant in the gospel of Christ, to strengthen you in your faith and to exhort you that no one be moved by these afflictions." To strengthen someone by a spiritual gift means to help their faith not give way as easily when trouble enters their life. We have spiritual gifts in order to help other people keep the faith and maintain an even keel in life's storms. If there is anybody around you whose faith is being threatened in any way at all, take stock whether you may have a spiritual gift peculiarly suited to strengthen that person. I think it would be fair to say also from this text that you shouldn't bend your mind too much trying to label your spiritual gift before you use it. That is, don't worry about whether you can point to prophecy or teaching or wisdom or knowledge or healing or miracles or mercy or administration, etc., and say, "That's mine." The way to think is this: The reason we have spiritual gifts is so that we can strengthen other people's faith; here is someone whose faith is in jeopardy; how can I help him? Then do or say what seems most helpful and if the person is helped then you may have discovered one of your gifts. If you warned him of the folly of his way and he repented, then perhaps you have the gift of "warning." If you took a walk with her and said you knew what she was going through and lifted her hope, then perhaps you have the gift of "empathy." If you had them over to your home when they were new and lonely, then perhaps you have the gift of "hospitality." We must not get hung up on naming our gifts. The thing to get hung up on is, "Are we doing what we can do to strengthen the faith of the people around us? I really believe that the problem of not knowing our spiritual gifts is not a basic problem. More basic is the problem of not desiring very much to strengthen other people's faith. Human nature is more prone to tear down than it is to build up. The path of least resistance leads to grumbling and criticism and gossip, and many there be that follow it. But the gate is narrow and the way is strewn with obstacles, which leads to edification and the strengthening of faith. So the basic problem is becoming the kind of person who wakes up in the morning, thanks God for our great salvation, and then says, "Lord, O how I want to strengthen people's faith today. Grant that at the end of this day somebody will be more confident of Your promises and more joyful in Your grace because I crossed his path." The reason I say becoming this kind of person is more basic than finding out your spiritual gift, is that when you become this kind of person the Holy Spirit will not let your longings go to waste. He will help you find ways to strengthen the faith of others and that will be the discovery of your gifts. So let's apply ourselves to becoming the kind of people more and more who long to strengthen each other's faith. Now, in Romans 1:12, Paul restates verse 11 in different words: I want to strengthen you with my spiritual gift, "that is, I want us to be mutually encouraged by each other's faith, both yours and mine." Paul does two things here. First, he uses the old "It's my pleasure" tactic. You remember my sermon on Christian Hedonism and humility? I argued that when we say, "It's my pleasure," after doing someone a favor, it is an expression of humility. It is like say, "Don't get too excited about my self-sacrifice; I'm just doing what I like to do." When Paul rereads Romans 1:11 he probably says, "Hmmm that may sound a bit presumptuous, as if I'm the great martyr doing all for their sake, when in fact I look forward to a great encouragement from them for myself." So as he restates verse 11 in verse 12 he adds that he, too, and not just they, is going to be helped when they meet. That is the first thing he does. The second thing he does is show that the way he will strengthen their faith by his spiritual gift (verse 11) is by encouraging them with his faith. In verse 11 he aims to strengthen them; in verse 12 he aim to encourage them. In verse 11 he strengthens faith by his spiritual gift; in verse 12 he encourages by his faith. The conclusion I draw from these parallels is this: a spiritual gift is an expression of faith which aims to strengthen faith. It is activated from faith in us and aims for faith in another. Another way to put it would be this: A spiritual gift is an ability given by the Holy Spirit to express our faith effectively (in word or deed) for the strengthening of someone else's faith. It is helpful to me to think about spiritual gifts in this way because it keeps me from simply equating them with natural abilities. Many unbelievers have great abilities in teaching and in administration, for example. And these abilities are God-given whether the people recognize this or not. But these would not be called "spiritual gifts" of teaching or administration because they are not expressions of faith and they are not aiming to strengthen faith. Our faith in the promises of God is the channel through which the Spirit flows on His way to strengthening the faith of others (Galatians 3:5).Therefore, no matter what abilities we have, if we are not relying on God and not aiming to help others rely on Him, then our ability is not a "spiritual gift." It is not "spiritual" because the Holy Spirit is not flowing through it from faith to faith. This has tremendous implications for how we choose church staff and church officers and board members (and Biblical Massage Therapists/Practitioners). It means that we will never simply ask, "who has the skill to be efficient?" We will always go beyond that and ask, "Do they use their skill in such a way that you can tell it is an expression of their hearty reliance on the Lord? And do they exercise their skill with a view to strengthening the faith and joy of others?" A church where the Holy Spirit is alive and powerful will be a church very sensitive to the difference between natural abilities and spiritual gifts. Now let's go on to Romans 12:3-8, a unit dealing in a bit more detail with spiritual gifts, though they are only called "gifts" here. "By the grace given to me I bid every one among you not to think of himself more highly than he ought to think, but to think with sober judgment, each according to the measure of faith which God has assigned him... (v.3). Having gifts that differ according to the grace given to us, let us use them: if prophecy, in proportion to our faith, if service, in our serving; he who teaches in his teaching; he who exhorts in his exhortation; he who contributes in liberality; he who gives aid with zeal; he who does acts of mercy, with cheerfulness (vv. 6-8).... Spiritual gifts are not a limited and defined group of activities spelled out in the New Testament. Rather, spiritual gifts are any ability the Spirit gives you to express your faith in order to strengthen another person. Notice the last four mentioned in verse 8,"exhorting" (or comforting, encouraging -- it's the same word used back in 1:12), "contributing" (or sharing), "giving aid" (may mean "presiding") and "acts of mercy." The remarkable thing about these (with the possible exception of "presiding") is that all believers are called to do these: exhort, give, and be merciful. So the "gift" must be that some are enabled by the Spirit to do it more heartily and effectively and frequently than others. So any virtue at all in the believer's life which he is enabled to do with zest and with benefit to others can be called his gift...And now, finally, turn to 1 Peter 4:10, 11, one of my favorite texts. I want to make four brief observations about spiritual gifts on the basis of these two verses. Let's read them. As each has received a gift, employ it for one another (or serve it up to one another) as good stewards of God's varied grace: whoever speaks, as one who utters oracles of God; whoever renders service, as one who renders it by the strength which God supplies; in order that in everything God may be glorified through Jesus Christ.

*****

If my people, which are called by my name, shall
humble themselves, and pray, and seek my face, and turn from their wicked ways; then will I hear from heaven, and will forgive their sin, and will heal their land. (2Ch 7:14)
*****

6
Love and Healing

"The relationship between an atmosphere of loving care and the process of healing brings into focus a critically important issue that has captured the attention of philosophers, theologians, and members of the medical profession throughout the ages," writes, Donald De Marco, Ph. D., professor emeritus of philosophy at St. Jerome's University in Ontario and an adjunct professor at Holy Apostles College and Seminary in Cromwell, Connecticut. "How is it possible for love, which is essentially spiritual, to have a transforming effect on the human body, which is corporeal and the natural object of scientific intervention? The answer to this timeless question, at least in part, lies in the integral wholeness of the human person. Love, as Pierre Teilhard de Chardin has described it, is the "affinity between being and being." In this sense, love is the great equalizer, having the inherent potential for expression between any one human being and another. Love, therefore, is an affirmation of the other, regarding the other in his wholeness. This affirmation rests on the recognition that a person's wholeness constitutes his original state, the state in which he is most himself. This original state is a person's fundamental good and, as such, is a natural object of love. None of us, needless to say, exists in a state of primal wholeness. Nor is any one of us unblemished. Consequently, the second phase of love is restoration. Here, love operates as the desire to help others return, as much as is possible, to that original state of wholeness. The simple act of a mother bathing her child and restoring that child to cleanliness exemplifies these two phases of love. By definition, restoration implies the original state which is the one that is affirmed. Restoration operates on a state that cannot be affirmed in itself. Love is intolerant of imperfection. It is important to note that in most European languages, the words health, healing, wholeness, and holiness are all etymologically related. Healing is a restoration of health and wholeness. We speak of physical health on a bodily level, mental health on a psychological level, integrity, referring to the moral dimension, and holiness, referring to spiritual wholeness. Healing is rooted in love insofar as love desires the other to be restored to wholeness, and this restoration process presupposes the primary and primal significance of wholeness. Disease, depression, sin, and alienation are all impediments that compromise wholeness. Healing involves the removal of these impediments...Christianity is monotheistic, and the love that is attributed to God is thoroughly divine. In fact, for Christians, God is identified as Love. Jesus, the Son of God, gave central importance to love and healing during his brief ministry on earth. Of the 3,779 verses in the four Gospels, 727, or nearly one fifth, are related specifically to the healing of physical and mental illnesses and to the resurrection of the dead. In most cases, Jesus combined the speaking of words with touching. His practice of touching began a tradition which survives in the present in the "laying on of hands." In three instances Jesus uses saliva in exercising his healing. Nor was his healing ministry constrained by social conventions. At least six of his healings took place on the Sabbath.

Writing about the great transition from the Greek to the Christian view of love and healing, J. W. Provonsha, M.D., had this to say: "It has become traditional to identify modern doctors in spirit with a long line of historic greats reaching back to the impressive Hippocrates. . .
But sometimes it is forgotten that medicine owes its greatest debt not to Hippocrates, but to Jesus. It was this humble Galilean who more than any other figure in history bequeathed to the healing arts their essential meaning and spirit. Physicians would do well to remind themselves that without His spirit, medicine degenerates into depersonalized methodology, and its ethical code becomes a mere legal system. Jesus brings to methods and codes the corrective of love without which true healing is rarely actually possible. The spiritual "Father of Medicine" was not Hippocrates of the Island of Cos, but Jesus of the town of Nazareth. Nonetheless, something of the philosophy of the ancient Greeks survives.

Writing from a Christian perspective, Morton T. Kelsey states in his book Healing and Christianity that "as this Spirit resides in us we build up defenses against alien forces so that they cannot so easily attack and possess us." Kelsey goes on to say that "love [is] an invitation to God's Spirit" which serves to protect us from "alien domination." He sees Christ's command to love one another as more than an ethical maxim. It also carries profound healing implications. The Christian approach to love and healing accepts that the wholeness of the person is of fundamental importance and that evil is something alien. In this regard, it owes much to ancient Greece. But it goes far beyond these ancients roots in that it holds love to be unequivocally divine, it expresses love in a more personal manner, and it understands the appropriateness of ministering directly to the body." The bible records 35 of them which include raising back to life three times people that died, 17 instances curing people with physiological disorders, including the blind, the lame, lepers and six times Jesus healed demoniacs. He healed by his word, sometimes by his touch.

Occasionally he used his saliva. "The blind receive sight and the lame walk, the lepers are cleansed and the deaf hear, the dead are raised up, and the poor have the Gospel preached to them." (Matt. 11:5)

7
What Is Meditation?

If your answer to this question has to do with the cliche about "emptying one's mind or concentrating on nothing," you have been listening too much to non-biblical philosophies. Turn to Joshua chapter 1 verse 8 KJV. "This book of the law shall not depart out of thy mouth; but thou shall meditate therein day and night, that thou mayest observe to do according to all that is written therein; for then thou shalt make thy way prosperous, and then thou shalt have good success." The book of the law shall not depart out of thy mouth." The book of the law to which Joshua was referring is the Pentateuch or the Five Books of Moses, Torah which is Hebrew for "the Law." However, to you and me, our "book of the Law" is our entire Word of God we call the Holy Bible, and Joshua is saying –keep on repeating the scriptures moment by moment, day and night. Study the word. Regurgitate the word. Reflect on the word. Then you will be conversant with God's Fasting Guidelines and His promises. Biblical meditation involves studying and regurgitating God's word, examining and reexamining it with a wholehearted reverential committed mindset to live by it. Apostle Paul puts it all together in his letter to the Philippians chapter 4 verses 6-8 in which he states, "Be careful about nothing; but in everything by prayer and supplication with thanksgiving let your requests be made known unto God. And the peace of God which passes all understanding, shall keep your hearts and minds through Christ Jesus. Now here comes the aspect of meditation." Verse 8 says, 'Whatsoever things are true, whatsoever things are honest, whatsoever things are just, whatsoever things are pure, whatsoever things are lovely, whatsoever things are of good report; if there be any virtue, and if there be any praise, think on these things." When we put our minds on the awesomeness of God. Continuously thinking about His Love and how we too as His children can emulate His Love, our minds will have no space for anger, grudge, malice, jealousy, gossip, foolish talking, jesting, lusting. We ourselves become nothing as our human Spirit reaches to align with the Holy Spirit. When the human Spirit aligns with the Holy Spirit, that is, being "in the will of God". Our Lord Jesus gave us this promise, "I am telling you the truth: whoever believes in me will do what I do – yes, he will do even greater things because I am going to the Father...I will do whatever you ask for in my name, so that the Father's glory will be shown through the Son." (John 14:12 – 13)

*****

My son, be attentive to my words
My son, be attentive to my words;
incline your ear to my sayings.
Let them not escape from your sight;
keep them within your heart.
For they are life to those who find them,
and healing to all their flesh.
Keep your heart with all vigilance,
for from it flow the springs of life.
  (Proverbs 4:20-23)
*****

8
Fasting and Prayers

The Lord Jesus

Mark 9: 23- 29 states, "If thou canst believe, all things are possible to him that believeth. And straightway the father of the child cried out, and said with tears, Lord, I believe, help thou mine unbelief. When Jesus saw that the people came running together, he rebuked the foul spirit, saying unto him, 'Thou dumb and deaf spirit, I charge thee, come out of him, and enter no more unto him.' And the spirit cried and rent him sore, and came out of him, and he was as one dead. But Jesus took him by the hand and lifted him up; and he arose. And when he was come into the house, his disciples asked him privately, 'Why could not we cast him out?' And he said unto them, "This kind can come forth by nothing, but prayer and fasting." Matthew 6:17-18 KJV states, 'When you fast, anoint your head and wash your face, that you appear not to men to fast, but in secret to your Father. Your Father, Who sees (you) in secret shall reward you openly. Account of an event narrated in Matthew 17:1-21, Luke 9.28-36 and Mark 9:2-29 shows our Lord Jesus while here on earth with His disciples taking Peter along with the two sons of Zebedee, James and John atop a high mountain where he rendezvous with Moses and Elijah. While Jesus was talking, "...a shining cloud came over them, and a voice from the cloud said, "This is my own dear Son, with whom I am pleased – listen to Him!" (Verse 5) When they returned to join the rest of the disciples down the mountain, they were surrounded by a great. A man came to Jesus, knelt before Him, and, "Sir, have mercy on my son! He is an epileptic and has such terrible attacks that he often falls in the fire or water. I brought him to your disciples, but they could not heal him." Our Lord called out the demon from the boy, and he was healed at that very moment. His baffled disciples later, asked Jesus how come they were unable to heal him. Jesus told them, 'Because of your unbelief: for verily I say unto you, if ye have faith as a grain of mustard seed, you shall say unto this mountain, remove hence, to yonder place; and it shall remove, and nothing shall be impossible unto you. Howbeit, this kind goeth not out, but by prayer and fasting. ('' Matt 17:1-21KJV) Translation: When you trust the power of God to do all things, even if the size of your faith is as the size of a grain of mustard seed, with God all things are possible, and because of that faith, you can even do the impossible as saying to a mountain to move out of here to another place. Even so, however, there are issues such as this type that can be addressed only by prayer and fasting.

On this account, three friends: a Seminarian, a farmer, and a cab driver embarked upon fasting to achieve some predetermined personal goals. The Seminarian fasted forty days and forty nights without food or water to pass an impending graduation examination while paying little attention to the scriptural advice, "Study to show yourself approved." (2 Timothy 2:15) The farmer fasted twenty-one days during the planting season, and never planted a seed. But the scripture warns, "He who does not work neither shall he eat. " (2Thessalonians 3:10) The cab driver fasted seven days, and went on a long journey with no gas in his car totally in violation of the command, " You must not put the LORD your GOD to the test." (Luke 4:12) God says, "My people are destroyed for lack of knowledge." (Hosea 4:6) Let us learn not to fast out of presumptuousness.

What Is Fasting?

So, what is fasting?

Fasting is not self-flagellation by means of voluntary starvation as a tool to force the hand of God to do our will, or a means to "attain a higher level of consciousness." Fasting is not involuntary hunger caused by total deprivation and want. Fasting is a determined effort to mortify the flesh in order to heighten spiritual sensitivity. Man, the bible says in Genesis 2:7 was made out of the dust of the earth, and God breathed in him the breath of life. And he became a living soul. On accepting the Lord Jesus as Savior the man or woman gets the indwelling of the Holy Spirit; in his or her regenerated new birth, s/he becomes Spirit, soul, and body. When we are saved, we become new bodies in Christ. Here is who the bible says we are, "When anyone is joined to Christ, he is a new being; the old is gone, the new has come." (2Corinthians 5:17) At time goes on, we become Spiritually rustic. When we get spiritually rustic, we fast to recharge our Spiritual batteries. Dr Adrian Rogers explains it in a way only he does so cogently. Have you counted the cost if your soul should be lost? Did you know that when God made you, God breathed into your nostrils

what the Bible calls, "the breath of life?" And you became, in the words of the Bible, "a living soul, made in the image of God." You are not a body; you have a body. The body is the house in which your soul is. The body will one day die. It will be put in the ground and it will decay. But your soul, made in the image of God will last for eternity. When the sun, the moon, the stars have fallen out of orbit, when the sun has become a cinder, when time shall be no more, your soul will be in existence somewhere, either in heaven or hell--timeless, dateless, measureless; your soul, made in the image of God will exist forever and ever and ever and ever. That's why the Lord Jesus Christ asked this question in the word of God. Mark 8:36, "What does a person gain if he wins the whole world and loses his life?" Fasting purposefully means being in prayer along with constant study of the word of God. Joshua expresses it succinctly when he echoed God's
command to the people of God in Joshua 1:8, "This book of the law shall not depart out of thy mouth; but thou shalt meditate therein day and night, that thou mayest observe to do according to all that is written therein: for then thou shalt make thy way prosperous, and then thou shalt have good success."

Biblical Fasting Guidelines

God has established guidelines for fasting. We
complain about our prayers not being answered
although we have fasted and prayed without ceasing because we fail to follow those guidelines.
This is what God said about "True Fasting". Open Isaiah 58: 1-14:
1 "Shout it aloud, do not hold back.
Raise your voice like a trumpet.
Declare to my people their rebellion
and to the house of Jacob their sins.
2 For day after day they seek me out;
they seem eager to know my ways,
as if they were a nation that does what is right
and has not forsaken the commands of its God.
They ask me for just decisions
and seem eager for God to come near them.
3 'Why have we fasted,' they say,
'and you have not seen it?
Why have we humbled ourselves, and you have not noticed?'

"Yet on the day of your fasting, you do as you please and exploit all your workers.
4 Your fasting ends in quarreling and strife,
and in striking each other with wicked fists.
You cannot fast as you do today
and expect your voice to be heard on high.
5 Is this the kind of fast I have chosen,
only a day for a man to humble himself?
Is it only for bowing one's head like a reed
and for lying on sackcloth and ashes?
Is that what you call a fast,
a day acceptable to the LORD?
6 "Is not this the kind of fasting I have chosen:
to loose the chains of injustice
and untie the cords of the yoke,
to set the oppressed free
and break every yoke?
7 Is it not to share your food with the hungry
and to provide the poor wanderer with shelter—
when you see the naked, to clothe him,
and not to turn away from your own flesh and blood?
8 Then your light will break forth like the dawn,
and your healing will quickly appear;
then your righteousness I will go before you,
and the glory of the Lord will be your rear guard.
9 Then you will call, and the Lord will answer;
you will cry for help, and he will say: Here am I.
"If you do away with the yoke of oppression,
with the pointing finger and malicious talk,
10 and if you spend yourselves in behalf of the hungry and satisfy the needs of the oppressed,
then your light will rise in the darkness,
and your night will become like the noonday.
11 The Lord will guide you always;
he will satisfy your needs in a sun-scorched land
and will strengthen your frame.
You will be like a well-watered garden,
like a spring whose waters never fail.

12 Your people will rebuild the ancient ruins
and will raise up the age-old foundations;
you will be called Repairer of Broken Walls,
Restorer of Streets with Dwellings.
13 "If you keep your feet from breaking the Sabbath and from doing as
you please on my holy day, if you call the Sabbath a delight
and the LORD's holy day honorable,
and if you honor it by not going your own way
and not doing as you please or speaking idle words,
14 then you will find your joy in the Lord,
and I will cause you to ride on the heights of the land and to feast on
the inheritance of your father Jacob."
The mouth of the Lord has spoken.

God has a conditional guideline for fasting. He tells us, "If you" ...
(Verse 13) "then you will find your joy...and I will cause you to ride on
the heights of the land..."

Fasting, meditation and prayers accompanied by
reading and studying the scriptures increase the
sensitivity of the human Spirit to become responsive to the promptings
of the Holy Spirit "Let the Spirit direct your lives, and you will not
satisfy the desires of the human nature. For what our human nature
wants is opposed to what the Spirit wants, and what the Spirit wants is
opposed to what our human nature wants. These two are enemies.
(Galatians 5:16-19) "To be controlled by the human nature results in
death; to be controlled by the Spirit results in life and peace." (Romans
8:6)

****

I have been hungry and thirsty.

"I have been in prison more times, I have been whipped much more, and I have been near death more often. Five times I was given thirty-nine lashes by the Jews; three times I was whipped by the Romans; and once I was stoned. I have been in danger from floods and from robbers, in danger from fellow Jews and from Gentiles; there have been dangers in the cities, dangers in the wilds, dangers on the high seas and dangers from false friends. There has been work and toil; often I have gone without sleep; I have been hungry and thirsty. I have often been without food, shelter and clothing..."

(2 Corinthians 11:24-27)

****

9
Exemplars of Fasting

Moses

Moses fasted 40 days and 40 nights without eating nor drinking water. (Exodus 34:28)

Purpose: To intercede for the nation of Israel for the sin of the people. "For I was afraid of the anger and hot displeasure with which the Lord was angry with you, to destroy you. But the Lord listened to me at that time also."

David

David fasted 7 days. (2nd Samuel 12:16-18)

Purpose: Fasting as an act of contrition and

repentance for one's sin. David fasted to plead God's forgiveness of his sins of covetousness and murder, and to intercede for his ailing child born to him by the wife of his general, Uriah whom David cuckolded. "The Lord caused the child that Uriah's wife had borne to David to become very sick. David prayed to God that the child would get well. He refused to eat anything, and every night he went into his room and spent the night lying on the floor... A week later the child died..."

King Jehoshaphat

King Jehoshaphat proclaimed a fast. (2 Chronicles 20;3- 4)

Purpose: To plea for help in time of trouble, and for guidance. "Jehoshaphat was frightened and prayed to the LORD for guidance. Then he gave orders for a fast to be observed throughout the country. From every city of Judah people hurried to Jerusalem to ask the LORD for guidance."

Ezra

Ezra called for a national fast (Ezra 8:21 KJV)

Purpose: To come under GOD'S protection from

enemies and plead His favor. "There by the Ahava Canal I gave orders for us all to fast and humble ourselves before GOD and to ask him to lead us in our journey and to protect us and our children and all our possessions.

Nehemiah

Nehemiah fasted. (Nehemiah 1: 4 -9)

Purpose: For success of a mission for God's glory. "For several days I mourned and did not eat. I prayed to God......Give me success today and make the emperor merciful to me."

Esther

(Hadassah) Esther called for 3 days of fasting without food nor water. (Esther 4: 15 -16)

Purpose: For Divine favor." Hold a fast and pray for me. Don't eat or drink anything for three days and nights."

Jesus To Disciples Jesus calls disciples to fast and pray. (Matt 17: 21)

Purpose: For spiritual strength and power to heal.

"However, this kind does not go out, except by prayer and Fasting."

Our Lord Jesus Jesus fasted 40 days and 40 nights. (Luke 4: 1)

"Jesus returned from the Jordan full of the Holy Spirit and was led by the Spirit into the desert, where He was tempted by the Devil for 40 days.

Purpose: For Spiritual strength to resist temptations by the Devil.

Cornelius

Cornelius prayed and fasted. (Acts 10:30 – 31)

Purpose: As an act of mortifying the body.

"Four days ago, I was fasting until this hour. At the ninth hour I prayed in my house. Behold, a man stood before me in bright clothing, and said, 'Cornelius, your prayer is heard. Your alms are remembered in the sight of God.'"

The church/The faithful/Ministers

The faithful fasted and prayed. (Acts 13: 1 -3)

Purpose: Fasting and prayer as an act of worship and ministry. "In the church at Antioch there were some prophets and teachers: Barnabas, Simeon (called the Black), Lucius (from Cyrene), Manaen (who had been brought up with Governor Herod), and Saul. While they were serving the LORD and fasting, the Holy Spirit said to them, Set apart for me Barnabas and Saul, to do the work to which I have called them." They fasted and prayed, placed their hands on them, and sent them off."

The Church, Paul and Barnabas fasted and prayed

Paul and Barnabas fasted and prayed before ordaining ministers.

Purpose: Fasting and prayer for the Holy Spirit to

48

release His anointing on the ministers of God in
Christ. For the anointing breaks the yoke.

"Paul and Barnabas preached the Good News...and won many disciples. They strengthened the believers and encouraged them to remain true to the faith... In each church they ordained elders, and with prayers and fasting they them to the LORD, in whom they put their trust. (Acts 14: 21 – 23)

*****

My son, be attentive to my words
My son, be attentive to my words;
incline your ear to my sayings.
Let them not escape from your sight;
keep them within your heart.
For they are life to those who find them,
and healing to all their flesh.
Keep your heart with all vigilance,
for from it flow the springs of life.
(Proverbs 4:20-23)

*****

10
How Long Should Fasting Last?

Let's say, a person, because of deprivation hasn't eaten any food or taken in any drink for a whole day. Does it mean that that person has fasted? Let us be a bit more specific; an earthquake victim in a harrowing life-threatening condition of entombment alive under the rubble of a Cathedral for seven days without food or water; would that person's abstinence from food and drink be considered as fasting? In these two scenarios one person went without food and water for a day, and another seven days. In spite of the length of time each one went without food and water; insofar as matters of the spirit are concerned, they were not fasting. All right. Suppose I decided to deprive myself of food and water for three days to bring down my weight. Would that be called fasting? No! None of the scenarios can be called fasting, despite the fact that both the condition and the length of time are common to fasting. Here's why!

Fasting is when we voluntarily deny our bodies the things they desire so that we can purposefully put God as center of our focus. "I beseech you therefore, brethren, by the mercies of God, that ye present your bodies a living sacrifice, holy, acceptable unto God, which is your reasonable service. And be not conformed to this world: but be ye transformed by the renewing of your mind, that ye may prove what is that good, and acceptable, and perfect will of God.

Translation by Today's English Version: "So then, my brothers, and sister, because of God's great mercy to us I appeal you: Offer yourselves as a living sacrifice to God, dedicated to His service and pleasing to Him. This is the true worship that you should offer. Do not conform yourselves to the standards of this world, but let God transform you inwardly by a complete change of your mind. Then you will be able to know the will of God –what is good and is pleasing to him and is perfect.

During fasting, we are aligning our bodies and souls to the Spirit of God within us to present ourselves as living sacrifices by denying ourselves the and desires of the body. We must close our minds to what the world says, or does, or hears, or sees. We must keep our minds refreshed by meditating on the WORD of God ceaselessly. Fasting and prayer are conjoined. We must pray without ceasing as the scripture commands us to do. When we do these things, here's God's promise: Then you will be able to know the will of GOD –what is good and is pleasing to him and is perfect.

To some, fasting one day is all that it takes to put the body under subjection to the Spirit; to another, seven days. To some, twenty-one days, to another 40 days. Cornelius fasted for one day for spiritual renewal and God answered him. Daniel fasted 21 days for spiritual understanding and had a delayed answer. Jesus fasted 40 days and 40 nights to overcome Satan's plots and the Devil fled from Him for a little while. Moses fasted 40 days to intercede for the people of Israel. God heard him. David fasted 7 days to petition God's forgiveness for his sins of adultery, perfidy and murder. God loved him. Our fasting must be purposeful. We must leave how long to fast to the leading of the Holy Spirit. And while we fast and pray, we must pray in the name of our LORD Jesus Christ. He promised: "Believe me when I say that I am in the Father and the Father is in me. If not, believe because of the things I do. I am telling you the truth: whoever believes in me will do what I do – yes, he will do even greater things, because I am going to the Father. And I will do whatever you ask for in my name, so that the Father's glory will be shown through the Son. If you ask me for anything in my name, I will do it. (John 14:11-14) We must stand on this promise of the Lord. He will empower us in a way that will baffle us. "Trust in the LORD with all thine heart; and lean not unto thine own understanding. In all thy ways acknowledge him, and he shall direct thy path." (Proverbs 3:5)

In The Will Of God

Asking God for anything in the name of Jesus is in the will of God. For our Lord Jesus has promised He will He will answer our prayers. We must not ask Almighty God for anything in the name of Jesus and expect an answer when we are habitually committing sin. Sin is not in the will of God. When we fast and pray, we must keep our purpose on God, and the will of God. James the brother of our Lord wrote this in his letter to "all God's people scattered over the whole world. "But when you pray, you must believe and not doubt at all. Whoever doubts is like a wave in the sea that is driven and blown about by the wind. A person like that, unable to make up his mind and undecided in all he does, must not think that he will receive anything from the Lord." (James 1:6 – 8) James also wrote this, "You do not have what you want because you do not ask God for it. And when you ask, you do not receive it, because your motives are bad; you ask for things to use for your own pleasures." (James 4: 2-3) Fasting and prayers must therefore be geared toward fulfilling the will of God. In 2 Samuel 12:16 -18, King David prayed to GOD that the child he had through adulterous relations with Uriah's wife, Bathsheba would get well. He refused to eat and every night he went into his room and spent the night lying on the floor. His court officials went to him and tried to make him get up, but he refused and would not eat anything with them. A week later, the child died.

Paul, in his 2 letter to Corinthians 12: 7 – 9, wrote that he had a painful physical ailment that acted as "Satan's messenger" to keep him from being puffed up with pride, "Three times I prayed to the LORD about this and asked him to take it away. But His answer was: My grace is all you need, for my power is greatest when you are weak."

Do not become dismayed when after all the fasting and prayers you have done, it would seem to you that God is not answering your prayers. Sometimes unanswered prayers turn out to be prayers answered. Look around you again and again. When you see that God has worked in a direction opposite to your prayers, you will also ultimately find out that all things have worked out for your good, and for His own glory. That is, fulfilling the will of God.

Always sing songs. Sing songs of praises. Praise the Lord at all times. Praising the Lord with songs of praises and worship break every yoke. Psalm 150 exhorts us to Praise the Lord. "Praise ye the Lord. Praise God in his sanctuary; praise him in the firmament of his power. Praise him for his mighty acts: praise him according to his excellent greatness. Praise him with the sound of the trumpet; praise him with the psaltery and harp. Praise him with the timbrel and dance: praise him with stringed instruments and organs. Praise him upon the loud cymbals; praise him upon the high-sounding cymbals. Let everything that hath breath praise the Lord."

11
Psalms and Songs

The prophet Samuel called David "a man after God's own heart" because David loved to pray and sing songs of praises and worship to the LORD. The Scripture says in Acts 13:22, "After removing Saul, he made David their king. God testified concerning him: 'I have found David son of Jesse, a man after my own heart; he will do everything I want him to do"

Note what God says of David: he will do everything I want him to do."

DAVID

1 Samuel 30: 6, "And David was greatly distressed, for the people spoke of stoning him, because all the people were bitter in soul, each for his sons and daughters. and every man his sons,

and for his daughters: but David strengthened himself in the LORD his God. It is safe to say a distinguishing characteristic of David is he loved to pray and sing Psalms of praises and worship to the LORD and was credited to be write some. In its introduction to the book of psalms, Good News Bible, Today's English Version, states, "The book of Psalms is the hymn book and prayer book of the Bible. Composed by different authors over a long period of time, these hymns and prayers were collected and used by the people of Israel in their worship, and eventually this collection was included in their Scriptures. These religious poems are of many kinds: there are hymns of praise and worship of God; prayers for help, protection, and salvation; for forgiveness; songs of thanksgiving for God's blessings; and petitions for the punishment of enemies. These prayers are both personal and national; some portray the most intimate feelings of all of one person, while others represent the needs and feelings of all the people of God. The Psalms were used by Jesus, quoted by the writers of the New Testament, and became the treasured book of worship of the Christian Church from its beginning." The Bible Dictionary states the following: "The ancient Hebrews called this collection Tehillim, that is, "songs of praise" or hymns. The fuller designation is Sefer Tehillim, that is, "The Book of Psalms." The expression, Psalm, is from Greek denoting "music on string instruments, or more generally, "songs adapted to such music." There are 150 psalms or 151 depending on whether you are reading from Protestants Bible or the one with the apocryphal/deuterocanonical books used in the Catholic Church. There are 73 psalms ascribed to David. Other psalms have been ascribed to Moses, Solomon, Heman , Ethan, Asaph, the sons of Korah and to anonymous writers. The 150 psalms have been divided into 5 groups since ancient time. 61 Psalms are written to be sung. Psalms 1- 41 ; 42-72 ; 73-89 ; 90 -106 ; 107-150. Of these 150 psalms, ten are believed to be specifically focused on petition for healing, according to traditional Hebraic sources. The ten healing Psalms are: 16, 32, 41, 42, 59,77, 90, 103, 105, 137 and 150. "Speak to one another with the words of psalms, hymns, and sacred songs; sing hymns and psalms to the Lord with praise in your hearts." (Ephesians 5:19).

Psalm 86 - One of my favorite Psalms when I need the favor of GOD, I fast to align my inner Spirit with the Spirit of God; and whenever I am unable to find words for what to say in prayers, or how to say it, I turn to Psalm 86 . If it was good for King David, it's good enough for me! It fully expresses my innermost thoughts and needs. So I keep my focus on its words while meditating on and mouthing them. Whenever I have the chance, I separate myself away from other people and pray this Psalm slowly, meaningfully, and reverentially. It has become my foundation prayer. Every day, I say this prayer to prepare Spiritually; that is , to calm down my personal anxieties, and have the frame of mind that equips me to intercede for others in prayers, knowing that Christ Jesus my Savior has them under His care. I end my prayer session with the words of Psalm 19:14.

Psalm 86 (New International Version)

A prayer of David.

1 Hear, O LORD, and answer me,
for I am poor and needy.
2 Guard my life, for I am devoted to you.
You are my God; save your servant
who trusts in you.
3 Have mercy on me, O Lord,
for I call to you all day long.
4 Bring joy to your servant,
for to you, O Lord,
I lift up my soul.
5 You are forgiving and good, O Lord,
abounding in love to all who call to you.

6 Hear my prayer, O LORD;
listen to my cry for mercy.
7 In the day of my trouble I will call to you,
for you will answer me.
8 Among the gods there is none like you, O Lord;
no deeds can compare with yours.
9 All the nations you have made
will come and worship before you, O Lord;
they will bring glory to your name.
10 For you are great and do marvelous deeds;
you alone are God.

11 Teach me your way, O LORD,
and I will walk in your truth;
give me an undivided heart,
that I may fear your name.
12 I will praise you, O Lord my God, with all my
heart;
I will glorify your name forever.
13 For great is your love toward me;
you have delivered me from the depths of the grave.
14 The arrogant are attacking me, O God;
a band of ruthless men seeks my life—
men without regard for you.
15 But you, O Lord, are a compassionate and
gracious God,
slow to anger, abounding in love and faithfulness.
16 Turn to me and have mercy on me;
grant your strength to your servant
and save the son of your maidservant.
17 Give me a sign of your goodness,
that my enemies may see it and be put to shame,
for you, O LORD, have helped me and comforted me.

Songs

Different moods of songs affect people in different
ways. Some songs stir up nostalgia, some deep
emotion, and some melancholy: some songs stimulate the imagination,
while others mirth and joy. Some songs are celebratory, some
devotional, and some praises and worship. Some songs produce a
feeling of calmness for the sick, or dying, some, a sense of relaxation
and comfort to one stressed down. Some songs can speed up the healing
process, relieve anxiety, and renew the spirit, soul and body. Moses
sang, and taught people to sing. (Ex 15. Deut. 32)

Israel sang songs along the journey to the Promised land. (Num. 21:17;
Judges 5) Deborah and Barak sang praises to God. David sang with all
his heart out. (Ps.104:33). Hezekiah's Singers Sang the Words of
David. (2 Chr. 29:28-30) The Lord Jesus and the Disciples sang at the
Last Supper. (Matt. 26:30)Jesus and the Disciples sang at the Last
Supper. Acts 16: 25; Paul and Silas sang in prison. In the dawn of
creation, "The morning stars sang together, and all the Sons of God
shouted for joy." (Job 38:7) In Heaven 10,000 times10,000 Angels sing
and the whole redeemed creation joins in the chorus. (Rev 5:11-13 ,
Moody Bible Handbook P231)

About "3000 years ago the servants of King Saul knew that the sounds
of the harp were able to have healing effect on people," writes Jewish
healing dot com. "They didn't have the scientific knowledge of why and
how, but they did know that something profound happened to people
when the harp was played, and so they found the 'best harpist' in the
land to bring to their ailing king to ease his suffering and bring him
moments of peace and wellbeing. 3000 years later, in our generation,
recent medical studies are revealing the 'whys and the hows' of the
healing effects of the sound of the harp."

Samuel 16; 14-23 (c1030-1010BC) says, "The Spirit of God departed
from Saul, and he was tormented by a spirit of sadness and melancholy
from God. Saul's servants said to him, 'Behold now! A spirit of
melancholy from God torments you. Let our lord tell your servants who
are before you that they should seek a man who knows how to play the
harp, so that when the spirit of melancholy from God is upon you, he
will play the harp with his hand and all will be well with you...So Saul

said to his servants, 'Seek now for me someone who plays well and bring him to me. One of the young servants spoke up and said, "Behold! I have seen a son of Jesse the Bethlehemite, who knows how to play the harp, is a mighty man of valor and a man of war, who understands a matter, is a handsome man, and God is with him....Saul sent messengers to Jesse, saying , "Send me David your son who is with the sheep....And it came to pass, that whenever the spirit of melancholy from God was upon Saul, David would take the harp and play it with his hand, and Saul would feel relieved and it would be well with him, and the evil spirit would depart from Saul."

The type of music to be played during Biblical Massage should be hymns or gospel songs specific to the treatment need of the client. Songs of praises for chair massage and needs requiring fast-paced techniques such as tapotement, cupping etc. Worship for effleurage etc relaxation. and medical massages.

****
Turn to me and have mercy on me;
grant your strength to your servant
and save the son of your maidservant.
Give me a sign of your goodness,
that my enemies may see it and be put to shame,
for you, O LORD, have helped me and comforted me.
****

12
Nutrition and Herbal Healing

I am including excerpts from my first book, God's Amazing Bible Plants Healed Me by way of telling my story attesting to God's surprises. I was not expected to live in 2004 because of a serious affliction. As my organs were not responding to the prescription medication, I remembered the promise of God in Ezekiel 47:12 that "plants are for food and for healing." I decided to search the scriptures for herbal treatment and pleaded with my doctor to monitor my progress or otherwise.

Here's my physician, Dr. Jon Hemstreet: "I remember well the first time I met Mr Scipio. He had been discharged from the hospital and presented to my office very ill. His symptoms involved almost every system in his body. Because of his multi-organ involvement, his prognosis was quite poor although no definite diagnosis could be established in the hospital. Mr Scipio and I discussed the next treatment steps we should take including additional testing, medications, and specialist consultation. Despite these further evaluations and more testing, his diagnosis remained elusive, and his condition did not improve. Mr Scipio stated on one follow-up visit that he wanted to be withdrawn from some of his medication and that he would pursue natural healing methods. Initially, I was not in favor of this approach, but eventually we agreed to keep him only on essential medications and monitor his progress with natural remedies. Remarkably, with time, Mr Scipio's condition improved to the point where his vital signs and laboratory values returned to normal. I was honored and surprised when Mr Scipio asked me to write the foreword to his book. I am trained in western allopathic medicine and do not have expertise in natural remedies. In fact, I am sometimes skeptical of these treatments and was not enthusiastic, in the case of Mr Scipio, to exchange his western medicine for natural ones. But his improvement has been a wonderful lesson for me in the power of natural substances to heal...

Jon E Hemstreet, M.D

13
Amazing Bible Healing Plants

2 Kings 20:1- 7 recounts, "About this time King Hezekiah became sick and almost died. The prophet Isaiah son of Amoz went to see him and said to him, The Lord tells you that you are to put everything in order, because you will not recover, Get ready to die." Hezekiah turned his face to the wall and prayed; Isaiah left the king, but before he had passed through the central courtyard of the palace the Lord told him to go back to say to him, "I, the Lord, the God of your ancestor David, have heard your prayer and seen your tears. I will heal you, and in three days you will go to the Temple. I will let you live fifteen years longer...Then Isaiah told the king's attendants to put on the boil a paste made of young leaves of fig tree and he would get well."

There are more than one hundred and twenty representative species of plants cited in the bible. God wants us to use the plants for food and for our healing, according Ezekiel 47:12. The fig that the prophet Isaiah instructed King Hezekiah to treat the illness has many medicinal uses; application of its latex on warts, skin ulcers, and sores produced positive outcome. The leaf tea was taken as a remedy for diabetes and calcification in the kidneys and liver.

Grapes, Vine (Vitis vini fera)
In Genesis 9:20, we are told that Noah who was
a farmer was the first man to plant a vineyard. Noah died at the age of 950 years. In a Time magazine article of February 1, 2010, Laura Blue wrote; "Elixir of youth sound fanciful, but the first crude anti-aging drugs may not be so far away. To date two compounds have sparked scientists' interests; resveratrol, a substance found in grapes, red wine and peanuts; and rapamycin... Resveratrol is currently available as a dietary supplement. A drug formulation of resveratrol is now being put to a more rigorous test, in clinical trials as treatment for Type 2 diabetes and cancers. In my book God's Amazing Bible Plants Healed Me first printed in 2007, I submitted as follows:
"Resveratrol may also be useful in controlling high levels of cholesterol, and in preventing the formation of blood clot. Grapes also help in fighting chronic inflammatory conditions such as arthritis and irritable bowels. Dr. Christopher Hobbs in his book, "Herbal Remedies For Dummies" wrote this: "The seed contain potent antioxidants

compounds called oligomeric proanthocyanidins (CPCs) which some researchers claim may slow down the aging process. Grapes come in many colors and sizes. The species has many varieties that are eaten as fruit, pressed for juice, dried as raisins or fermented into wine for drinking, flavoring food, for medicinal and religious use. The most popular varieties are the seedless. Grape juice is evaporated into grape honey. Pressed grape residue is used to make cream of tartar.

Grape leaves used as food wrap.

Pressed grape seeds yield a light vitamin rich fine culinary oil. it is ideal for aromatherapy massage. Grapes are useful help in restoring energy in convalescence. They are the basis of some blood cleansing diets. Chewing the seeds may stimulate anti-carcinogenic activity. The seeds of purple grapes contain powerful antioxidant compounds known as oligomeric proanthocyanins (OPC) which are

claimed to slow down the aging process. Grapes also help in fighting chronic inflammatory conditions such as arthritis and irritable bowels. The branch sap provides eyewash. Wine in moderation is regarded as a tonic, and readily absorbs properties of steeped herbs, so does grapes smoothies.

Medicinal Uses

Grapes are useful help in restoring energy in convalescence. In some quarters, medical experts have claimed that imbibing a moderate amount of wine relaxes muscles and mood, expands blood

vessels to lower blood pressure, and temporarily lower risk of heart diseases either by reducing stickiness of blood platelets, or by relaxing blood vessels, making them temporarily larger, or by increasing amount of HDLs." Wine is an antiseptic, sedative, analgesic – all three reasons for its use by "the Good Samaritan". (Luke 10:25-37) Ellagic acid occurs in grapes and may have a number of human health effects. It has anti-cancer properties and may act as free radicals scavenger.

The branch sap : Provides eyewash.

Leaf: Is used as tea to treat diarrhea, hepatitis, thrush and stomachache.

Leaf poultice: Applied externally for sore chest,

headache, rheumatism, and fevers.

Grapeseed: Varieties containing seeds when chewed and swallowed with the sweet pulp are believed to help in protecting one's internal organs against stress and environmental toxins. Chewing the seeds may stimulate anticarcinogenic activity. The seeds of purple grapes contain powerful antioxidant compounds known as oligomeric proanthocyanins (OPC) which are claimed to slow down the aging process.

Juice: Purple grape juice contains the same powerful disease-fighting antioxidants called flavonoids that are believed to give wine many of the heart-friendly benefits.

Wine: It contains antiseptic, sedative and analgesic properties.

Food

Grapes are used to produce raisins for condiments,

sweeteners in bread, cakes and pastries. Grapes are used to produce juices, wines and vinegars for

culinary preparations. Young leaves infusions are

used as beverages and food additives.

Fish and Grape Leaves in Phyllo Recipes

Ingredients

¾ lbs sliced mushrooms

2 cloves garlic, minced

3 cloves of raw garlic paste

2 tsp olive oil

4 large pickled grape leaves

¼ tsp crushed dried mint

¼ tsp crushed dried thyme

¼ tsp crushed dried sage

8 sheet frozen phyllo dough

1 cup grapes smoothie

2 medium size whole fish of choice, e.g, salmon, cleaned, skinned, deboned and halved lengthwise.

Instruction

Cook mushrooms and garlic in oil 4 to 5 minutes or until tender, then set aside. Remove and discard stems from grape leaves, and rinse thoroughly. Stack and roll up leaves as for jelly roll. Cut into 1/8" slices, then crosswise into pieces. Mix in bowl mushroom, garlic mixed with grape leaves, mint, thyme and sage.

Take 4 phyllo sheets. Mix grapes smoothie, and vinegar with garlic paste into a fine dressing consistency.

Brush sheets of phyllo with a liberal amount of garlic-grapes smoothie dressing, then stack up sheets. Cut stack in half, forming 2 (14"x19") rectangles. Spread about 1and 1/2 tablespoons of dressing on each side of fish. Place fish in 1 corner of phyllo sheets. Fold corner over, then sides over and roll up to form a package. Place in a clean baking dish. Repeat with remaining fish. Bake at 375 degrees for 20 to 25 minutes or until golden.

Honey

Apis Mellifica

There are more than 20,000 species of bees, but only 5 kinds produce honey. To make one pound of honey,160,000 honeybees would make as many as two million trips to flowers so that they could collect four pounds of nectar to produce honey one fourth the quantity of nectar collected. One honeybee would take an entire lifetime to produce a teaspoon of honey.

Constituents

Honey is the only food that will not spoil. It may crystallize in cold weather or over a time, but will melt under heat. It is a source of rapid energy. It contains live enzymes vital for the proper functioning of our body systems. Its other constituents include glucose, fructose, proteins, antimicrobials, hormones, organic acids, and carbohydrates. Honey is composed of a complex mixture of vitamins, namely; vitamin B6, vitamin B-12, vitamin C, vitamin A, vitamin D, vitamin E, and vitamin K as well as thiamin, riboflavin, niacin, foliate, and pantothenic acid. Honey is rich in such minerals as calcium, copper, iron, magnesium, Manganese, phosphorous, potassium, sodium, and zinc. Honey as a simple sugar, breaks down easily requiring the body less work in converting the sugar to energy, unlike refined white sugar which has been proven to be carcinogenic the sugar to energy, unlike refined white sugar which has been proven to be carcinogenic, inhibits calcium absorption in the intestines and weakens the strength ability of the vitamins and minerals that honey has put in the body.

Medicinal Uses

Honey is an antibiotic, antiviral, anti-inflammatory, anti-anemic, tonic, laxative, anti-allergenic, expectorant, and anti-carcinogenic. It is an overall tonic to all systems in the body and is of special use in the intestinal and skeletal systems. Because of its anti-microbial properties honey is effective in healing wounds. It has proved useful in burns and sunburn. The glucose in honey speeds up the body's absorption abilities of calcium, zinc, and magnesium, thus, providing a quick energy boost. Honey mixed with vinegar can soothe arthritic joints. Honey has been used throughout the ages to heal wounds. When honey mixes with the fluids from the body in the wound, it actually, causes those cells to release hydrogen peroxide to cleanse the wound and promote healing. Honey has also been proven useful in healing ulcers and gastric lesions. Its specific properties have proven beneficial in treating respiratory ailments. It is best to purchase raw, unmolested honey to receive the maximum benefits possible from the honey.

Food

If you're a consumer of honey, don't be too sure you have the best honey full of needed nutrients, if your honey is clear and fluid. The absence of your honey's cloudy appearance is proof that the honey has been strained thoroughly through a fine sieve, thus removing from it compounds containing valuable pollen nutrients including vitamins and minerals that body needs for its proper functioning. What you're getting from this type of fancy honey is a sweetener. Processed honey comes in four classifications; Grade A (Choice), Grade B (Standard), Grade C (Standard),

Grade D (Substandard).

Honey is about 140 percent sweeter than sugar. Use less amount of it than you would refine white sugar. Darker honeys are generally richer in nutrients than light honeys. The best honey will congeal in a room temperature. It is called uncooked honey when the degree of heat in melting down the honey is less than 104 degrees Fahrenheit, thus preserving the component enzymes, vitamins and minerals. When honey is highly heated its chemistry alters.

Ingredients

1 Gallon spring water

8 Table spoonful or bags Balm (Melissa) / Chamomile will do just fine

16 Ounces Raw Dark honey

16 Ounces Fresh ginger roots.

Instructions:

Peel fresh ginger roots

Put ginger roots in blender and process to a fine paste

Remove ginger paste from blender and put in vessel ready to heat with water.

Add gallon of water and stir

Add fresh or dried balm leaves

Put on stove bring to boil under low 80 degrees

Fahrenheit temperature

Add honey and stir for 5 minutes to mix

Remove from stove and cover to steep for 20 minutes.

Strain

Bottle infusion and refrigerate

Enjoy as desired before bedtime

Olive

Constituents:

Gum-resin, benzoic acid, olivile, mannite oleuropein, oleuropeoside, arachidic esters, free Oleic Acid, Oleum palmitic acid, palitoleic acid, steric acid, oleic acid, linolenic acid, linoleic acid, hydrocarbon, squalen, sterols, carotinoids, tocopherol.

Medicinal Parts Used:

Oil of the fruit, leaves, bark.

Medicinal Use.

Olives are used as febrifuge, antibacterial, antifungal, antiseptic, antiviral, astringent, tranquilizer, aperients, cholagogue, and as an emollient. They are also used to stabilize blood sugar levels, and in treating viral or parasitic conditions such as giardia, intestinal worms, malaria forming protozoa, microscopic protozoa pinworms, ringworm, roundworm, and tapeworms.

Olive leaves are used as an antiseptic. Olive leaf infusion has been used in treating hypertension, because of its capacity to lower blood pressure, and inhibit oxidation of LDL ("bad") cholesterol. When heated olives are brined to preserve them, oleuropin is converted into another chemical called elenoic acid.

Lenolic acid has shown antibacterial actions against several species of Lactobacilli and Staphylococcus aureaus and Bacillus subtilus in a test tube study.

Olive oil

Olive oil contains monosaturated fats, primarily, oleic acid, squalene, and phenolic compounds that function as antioxidants in the body. Oleuropein, responsible for the bitterness of raw olives, is one of the phenolics. Other simple phenols including tyrosol, lignans and pinoresinol also function as antioxidants.

Extra Virgin Oils are higher in these protective

compounds than processed oils. Olive oil may act by reducing the LDL ("bad") and raising the HDL ("good") forms of cholesterol in the blood. Olive extracts have been shown to have hypoglycemic activity, and oil reduces gallstone formation by activating the secretion of bile from the pancreas.

Olive oil may act as a mild laxative. The inferior oil is used for soap.

Food

Eating Bible Land Foods so-called "Mediterranean Diet", rich in olive oil, fruits, vegetables, and fish, according to nutritionists, lower the risk of colon, breast, and skin cancer as well as coronary heart disease. Olives are picked green when they are unripe, or black when ripe. Ripe olives are repeatedly pressed for olive oil. Cold pressed virgin oil is the best quality for culinary consumption. It has high antioxidant content, and it is free from cholesterol. It is valued in salad oils and a good source of edible oil. Olives are used in food preservation.

Olive Tea

To make a tea, steep one teaspoon (5 grams) of dried leaves in one cup (250ml) of hot water for 10-15 minutes. Drinking olive leaf tea is beneficial to diabetic patients as it lowers blood sugar level, according to medical experts.

Garlic and Olive Focaccini--Diabetic Version

Ingredient

1 lb. Prepared pizza dough, divided into 4 (4 oz.) balls

2 Tbsp. pureed roasted garlic

1 cup Olives, halved

1/2 cup shredded asiago cheese

2 tsp. Chopped rosemary

2 tsp. Chopped thyme

Black pepper, to taste

Instruction:

Shape each piece of pizza dough into a 5-6 inch disc.

Place in a well-greased baking sheet. Spread 1/2 tablespoon of garlic puree on each, then dot with ¼ cup of Ripe Olives. Sprinkle with 2 tablespoons of asiago cheese, 1/2 teaspoon of rosemary and thyme .

Then add black pepper to taste. Allow to rest in a warm place for 30-60 minutes, then bake in a 400 degree oven for 15-20 minutes until lightly golden.

Serves 4. Serving Suggestion: Cut into 1-inch thick strips before plating.

Dijon Sautéed Chicken--Diabetic Version

Ingredient

4 (5 oz. each) Boneless, skinless chicken breasts, pulverized thin.

Add Kosher salt and coarsely ground black pepper to taste (optional).

1 Tbsp. All-purpose flour

2 Tbsp. Olive oil

1 (8 oz.) pkg. Frozen artichoke hearts

1/2 cup White wine

1/2 cup

Low sodium chicken broth

2 Tbsp. Stone ground Dijon mustard

2 Tbsp. Chopped tarragon

1 cup Ripe Olives, halved

Instruction:

Season chicken breasts with salt and pepper to taste, then sprinkle with flour. Heat 1 tablespoon of oil in a large high-sided sauté pan over medium-high heat.

Place chicken breasts in pan and cook for 3-4 minutes on each side until golden brown and cooked through.

Transfer to a clean plate and set aside. Pour remaining oil into pan and heat. Carefully, add artichokes and cook over medium heat for 2-3 minutes, stirring occasionally until golden. Whisk in red grapes paste, chicken broth, mustard and tarragon. Pour in Ripe Olives and return chicken to pan. Cook until heated through. Serve immediately.

Creamy Dijon Mustard. Makes 10 two table-spoon

serving

Ingredients

4 ounces soft plain tofu

¼ cup Dijon mustard

2 Tablespoons balsamic vinegar

1 tablespoon capers

1 teaspoon fresh tarragon or ¼ teaspoon dried

tarragon

2 6-inch scallions, minced

1 tablespoon chopped flat-leaf parsley

½ cup Homemade Vanilla Glucema Shake

Dash of Tabasco sauce

Freshly ground black pepper, to taste
Instruction
Place the tofu in the bowl of food processor and
puree.
Add all remaining ingredients and process until
smooth.
Store the dressing in an airtight container in the
refrigerator until serving time. The dressing can be
stored for one day after it has been prepared.
Serve with 1 cup of raw, assorted vegetables such as colored peppers,
spears of endive, asparagus, fresh grape leaf shoots, celery sticks, fresh
fennel.

Balm
Also known as: Melissa Officinalis, Lemon Balm,
Sweet Balm., Bee Balm, Balm Mint, Blue balm,
Cure-All, Dropsy Plant, and Garden Balm.
In Genesis 43:11, the patriarch, Jacob, is quoted as
instructing his son to take as gifts "…a little balm…" among others to
"the man" in Egypt.
Lemon Balm was believed to be an elixir of youth. It was used as part
of a drink to ensure longevity.
The name Melissa is from the Greek word for "bee" officinalis,
meaning 'workshop' an indication of the compelling attraction the
plant has for honey bees.
The word Balm is short form for Balsamon, referring to the plant's oily
fragrant resin credited with the ability to soothe and calm nerves.
Constituents:
Volatile oil including citronella, polyphenols, tannins, flavonoids. It is
carminative, diaphoretic and febrifuge. Balm is also used as an
antidepressant, sedative, antiviral, antibacterial, and antispasmodic
ointment.
Medicinal Uses
It induces mild perspiration and makes a refreshing tea to treat early
stages of cold, fever, catarrh, and influenza. Balm is a useful herb,
either alone or in combination with others. It is excellent for use in
colds attended with fever, as it promotes perspiration.
It is also taken for depression, nervous exhaustion,
sleeplessness, menstrual cramps, nausea, and

indigestion. It repels mosquitoes.

Lemon Balm may be used alone by itself or in combination with other herbs. Used externally, lemon balm may relieve painful gout swellings. Harvest before flowering.

For a relaxing Lemon balm bath, tie up a bunch of fresh or dried lemon balm leaves in a towel or fabric gauze. Place in bathtub. Run hot water over. Enjoy the refreshing lemon-mint health-giving aroma.

For treating minor wounds and insect bites, make a hot compress putting 4 crushed tablespoon of lemon balm in a cup of water. Boil for 10 minutes. Soak a clean towel in the solution and place it on the wound.

For treating herpes simplex, make a poultice of fresh lemon balm. Add a teaspoon of dark, uncooked honey. Apply ointment on herpes lesion. Same lemon balm and dark uncooked honey ointment applied on sore, itching groin will speed up healing.

For repelling mosquitoes and other insects, rub crushed fresh lemon balm leaves or balm oil on exposed body parts.

For calming nerves, restful sleep, reducing fever or easing menstrual cramps, treating anxiety, indigestion and acidity due to eating too much, too quickly or missing meals, bloating, heartburn or stomach pain.

Food

It makes a refreshing replacement for carbonated soft drinks. Lemon-mint scented, honey sweet balm is used also as flavoring to salad and salad dressing, or garnish in soups, stews, custards, pudding or cooking.

Lemon Balm Tea

Preparation:

Take 2 Tablespoon fresh lemon balm leaves

2 Tablespoon or 4 bags Chamomile

1-pint fresh water

Honey

Instruction:

Boil fresh lemon leaves and chamomile in water for 10 minutes. Pour infusion in a cup.

Add honey as desired

Take at bedtime

Lemon Balm Drink

Ingredients
1 Gallon spring water
8 Table spoonful or bags Balm (Melissa) / Chamomile
will do just fine
16 Ounces Raw Dark honey
16 ounces Fresh ginger roots.
Instructions:
Peel fresh ginger roots
Put ginger roots in blender and put in vessel ready to heat with water.
Add gallon of water and stir
Add fresh or dried balm leaves
Put on stove bring to boil under low 80 degrees
Fahrenheit temperature
Add honey and stir for 5 minutes to mix
Remove from stove and cover to steep for 20 minutes.
Strain.
Bottle infusion and refrigerate
Enjoy as desired before bedtime
Read a chapter of the Word of God, meditate on the Word, and pray as you invite Jesus Christ into your heart. Your life will never be the same again! Amen

*****

I pray that all may go well with you
"Beloved, I pray that all may go well with you and
that you may be in good health, as it goes well with your soul."
(3 John 1:2)

*****

14
Biblical Massage Therapists

As Biblical Massage Therapists, we must look to Almighty God for His healing anointing that you can serve others for His glory through Christ Jesus. The Lord Jesus himself emphasized the importance of believing. In recent years science is beginning to acknowledge the role of belief in restoring health. According to David H. Rosmarin,PhD, McLean Hospital clinician and instructor in the department of Psychiatry Harvard Medical School, patients who expressed their faith in God, and an expectation of Divine intervention in their recovery, experienced a much shorter time of healing than the patients who did not believe in God. A Biblical Massage Therapist/Minister, witnessing for Christ must therefore "...hold fast the profession of our hope." (Heb 10:23) Be ready in season and out of season to preach the gospel of salvation by the Lord Jesus Christ ; tell about His immaculate conception, his birth; tell the story of his life, his teachings, about how and why he was crucified, how he died, was buried, and how he rose from death and ascended to heaven ; tell of our expectation that he is coming back again and our hope that believers now dead will rise again to meet him face to face. Jesus is telling you and me, "Look, I am sending you out as sheep among wolves. So be as shrewd as snakes and harmless as doves." (Matthew10:1) Avoid the slightest appearance of syncretic practices because the God we serve, Himself has said He is a jealous God. "I Am Yahweh, that is my name, and my glory I will not give to another; neither my praises to graven images. (Isaiah 42:8) A Biblical Massage Therapist must exercise care that while desiring an eclectic practice he or she is not lured by seemingly innocent but idolatrous modalities in any form or guise. Reject reiki. It is sorcery. Sorcery involves casting spells. Sorcery is about manipulating "energy," another name for principalities, cosmic powers of this present darkness, and spiritual forces of evil in the heavenly places (Eph 6: 10-12). Thai massage reverences Buddhist's gods. Tai chi pays homage to the dragon. Yoga has its roots in Hinduism.

With Biblical Massage therapy, because the power of the Holy Spirit will be working through the practitioner to heal the sick, the practitioner will have no need to apply any technique that is the product of polytheistic or new age tradition. Says the scripture, "But as for you, Christ has poured out His Spirit on you. As long as His Spirit remains in you, you do not need anyone to teach you. For His Spirit teaches you about everything, and what He teaches is true, not false. Obey the Spirit's teaching, then, and remain in union with Christ." (1 John 2: 26 -27) The notion that "energy" is the "source" of healing and therefore can answer for the god of any religion in any culture according to the perception and inclination of each individual is syncretic. It is a lukewarm blend of all belief systems, while feigning affinity with Biblical Christian dogma. Christ holds in contempt lukewarm believers. The Minister/Practitioner of Biblical Massage must regard as anathema to Biblical/ Christian World View of Healing any modality that shies away from acknowledging Almighty God through Christ as the One Who heals. Ayurvedic principles, for example, defy practices forbidden by God's first and second commandments, "Thou shalt have no gods before me." (Ex. 20:1-3)

"Hear O Israel! The Lord is our God, the Lord is One.
And you shall love the Lord your God will all your heart and with all your soul and with all your might." (Deut 6:4-5; Deut. 10:12; Num.11:18; Mark 12"29-30; Matt22:37 & Luke 10:27)

"You shall not make for yourself an idol, or any likeness of what is in heaven above or on earth beneath, or in the water under the earth. You shall not worship them, or serve them, for I, the LORD your GOD am a jealous GOD..." (Ex.20:4-5)

"The Lord God is supreme over all gods and over all powers. He is great and mighty, and he is to be obeyed. He does not show partiality, and he does not accept bribes. Have reverence for the LORD your GOD and worship Him only. (Deuteronomy 10: 17-20)

"...The true worshipers shall worship the Father in spirit and truth; for such people the Father to be His worshipers. (John 4:23) "...GOD is spirit and those who worship HIM must worship HIM in spirit and truth..." (John 4:24)

15
Christians, Reiki, and Ayurveda

Ayurveda claims there are four reasons why
man exists, namely,
(1) "Dharma" that is, to follow a destiny, a purpose or career.
(2) "Artha," that is, material accumulation.
(3) "Kama", pleasure.
(4) "Moksha," that is, liberation.

The definition of Ayurvedic "existence" from biblical viewpoint implies a state of superficial or temporal survival as opposed to "living" in Christ which is GOD'S gift of life in eternity to everyone who believes in His Son Jesus Christ. Those that believe in Jesus Christ do not exist. They live. They live because Christ lives. The reason we live is because the Lord Jesus lives in us, and we live to please God. The reason we live is to do the Will of God. The Will of God is borne out of God's Love for humanity, which is the reason
Almighty God gave His Only Son to be crucified so that whoever believes in Him will not die but live forever. For God is Love. Believing in the Son of God is expressed through acts of unqualified Love. Unqualified Love expresses God. We know this from 1 Corinthians 13: 4-7,
"Love is patient and kind; it is not jealous or conceited or proud; love is not ill-mannered or selfish or irritable; love does not keep a record of wrongs, love is not happy with evil, but is happy with the truth. Love never gives up; and its faith, hope patience never fail. Love is eternal..."
We express Love by our acts of Service, Selflessness and Sacrifice which fly in the face of Ayurvedic artha's material accumulation. Love heals.
As Ministers, Therapists, the scripture urges us, therefore, by the mercies of God, to present our "bodies a living and holy sacrifice, acceptable to God, which is your spiritual service of worship. And do not be conformed to this world, but be transformed by the renewing of your mind, that you may prove what the Will of God is, that which is good and acceptable and perfect."(Rom. 12:1-2) God calls Biblical Massage Ministers, Therapists to "...Walk in the Spirit, and ye shall not fulfill the lust of the flesh...But the fruit of the Spirit is love, joy, peace,

long-suffering, gentleness, goodness, faith, meekness, temperance against such there is no law...If we live in the Spirit, let us also walk in the Spirit." (Galatians 5:16 -25) The goal for individual pursuit of material prosperity set forth in Ayurvedic traditions is shared by those who preach "prosperity gospel" with effusive abandon from many televangelical mega-church pulpits. But that is far from biblical Christianity even though "prosperity gospel" preachers belabor their point that if money had been an issue with our Lord Jesus Christ, he would not have had any need for a treasurer, and Judas Iscariot would not be appointed treasure for him keep stealing money for the group without being found out quickly. Sometimes for such unrewarding arguments, it is best to "give a dog a long rope to hang himself." Enough said about that. Jesus the God-man who had the whole world in his hands, and who ordered Peter to go and catch a fish, open its mouth, find money in its mouth so Peter would use it to pay taxes for both himself and the Lord Jesus, this Jesus never encouraged his disciples to indulge in material accumulation on earth. Instead, he told them, "Do not store up riches for yourselves here on earth, where moths and rust destroy, and robbers break in and steal. For your heart will always be where your riches are. (Matthew 6:19-22) Any religion, ideology, or principle that bases its belief on material accumulation is teaching greed. Here is what Jesus said again about material accumulation, "If anyone wants to come with Me, he must forget himself, carry his cross, and follow Me. For whoever wants to save his own life will lose it; but whosoever loses his life for my sake will find it. "Will a person gain anything if he wins the whole world but loses his life?" (Matthew 16:24-26) As if on cue, Paul said this about his own life, "I have been in prison more times, I have been whipped much more, and I have been near death more often. Five times I was given thirty-nine lashes by the Jews; three times I was whipped by the Romans; and once I was stoned. I have been in danger from floods and from robbers, in danger from fellow Jews and from Gentiles; there have been dangers in the cities, dangers in the wilds, dangers on the high seas and dangers from false friends. There has been work and toil; often I have gone without sleep; I have been hungry and thirsty. I have often been without food, shelter and clothing..."
(2 Corinthians 11:24-27)
Does this account show Paul, the man who wrote two thirds of the New

Testament as a person enjoying accumulated material prosperity or, as one preaching "prosperity gospel"? Yet he firmly declared, "Who then can separate us from the love of Christ? Can trouble do it, or hardships or persecution or hunger or poverty or danger or death?...For I am certain that nothing can separate us from his love; neither death nor life, neither angels nor heavenly rulers or powers , neither the present nor the future, neither the world above nor the world below – there is nothing in all creation that will ever be able to separate us from the love of God which is ours through Christ Jesus our Lord." (Romans 8:24-39)

"Prosperity gospel" preached from the pulpits or as Ayurvedic artha for material accumulation, by any other name, is still antithesis of Biblical Christian way of life. That is why, in his letter to Timothy, apostle Paul wrote in 1 Timothy 6:10 , "The love of money is the root of all kinds of evil, and in their eagerness to be rich some have wandered away from the faith and pierced themselves with many pains." He also wrote this, "those who live as their human nature tells them to, have their minds controlled by what human nature wants. Those who live as the Spirit tells them to, have their minds controlled by what the Spirit wants. To be controlled by the human nature results in death; to be controlled by the Spirit results in life and peace." (Romans 8:5-6)

As Biblical Massage Therapists, our goal must not be toward material accumulation. That is what the bible calls greed, and , "greed is idolatry" (Ephesians 5:5) Healing, that comes from the touch ; the Biblical Massage Practitioner/Minister/ Therapist, must be grounded in Love which is expressed through service, selflessness and sacrifice. When God-Love is transmitted through our touch, a supernatural activity is innervated within the God-Spirit of the Toucher/Therapist and the God-Spirit of the Touched/Client/Sick. The resulting interaction is Spiritual harmony, releasing health, and wellness to the sick person's Spirit, soul, and body.

Reiki

Reject reiki. It is sorcery. Sorcery involves casting spells. Sorcery involves manipulating "energy," another name for cosmic powers of this present darkness, spiritual forces of evil in the heavenly places, principalities. Believers! The Spirit of the LORD exhorts you "to take up the whole armor of God so that you may be able to withstand on that evil day, and having done everything, to stand firm." (Eph.6:10-18). We know from 2 Corinthians 10:3-5, that, ""We live in this world, but we don't act like its people, or fight our battles with the weapons of this world. Instead, we use God's power that can destroy fortresses. We destroy arguments, 5 and every bit of pride that keeps anyone from knowing God. We capture people's thoughts and make them obey Christ. " (2 Cor.10:3-5)

16
God's Healing Touch

The gospel of John 11:38-45 reports that Lazarus had been dead for four days and was in a state of decomposition. Jesus raised Lazarus back to life!

The dead have no innate force to will themselves to rise. In our day, people who have been medically certified as dead, have come back to life, and have doctors baffled: a pregnant mother went to the delivery room on Christmas Day, 2009, according to CNN. She had complications, and was pronounced dead, along with her baby. Hours later, the medical personnel saw life coming back into the baby, and next, the mother. They lived. They were D-E-A-D.

They had no innate force. They were recipients of the gift of life from our Sovereign Lord, and Creator, who exercises His Sovereign Right to give His grace to whomever He Wills. So, the actions of the Practitioner are means of ministration or service to demonstrate God's Love for humanity, and for the sick/client that he/she may recover.

The miracle of healing is one of the means by which Almighty God glorifies Himself.

Matthew 8:2-3 reports, "A man with leprosy came and knelt before him (JESUS) and said, "Lord, if you are willing, you can make me clean." Jesus reached out his hand and touched the man. "I am willing," he said. "Be clean!" Immediately he was cured of his leprosy."

This man with leprosy exemplifies those willing to surrender to the authority of Yahweh Rapha, Jehovah God Who heals. Here is another report: "A man had been ill for thirty-eight years. When Jesus saw him lying there and learned that he had been in this condition for a long time, he asked him, "Do you want to get well?" (John 5:5-6)

The sick must not give up on life. He or she must show the willingness to get well. He or she must trust in God to get him or her well. He or she must believe the Biblical Massage Therapist/ Minister is acting upon authority from Christ Jesus to lay hands upon him or her and he or she will be healed. This is the authority the Minister/Therapist stands upon, according to the gospel of Mark 16:15-18, "Go in to all the world and preach the gospel to all creation. He who has believed and has been baptized shall be saved; but he who has disbelieved shall be condemned. And these signs will accompany those who have believed: in My name they will cast out demons, they will speak with new tongues, they will lay hands on the sick, and they will recover..."

Client/Patient

The Action of the Client / sick / recipient is to

surrender to the Will of Almighty God and be open to receiving the free gift of healing which is given by Almighty God's grace through faith in Christ Jesus. In other words, God our Sovereign Lord has already decided before the foundation of the world that He is bestowing His grace of healing on this particular sick person at this specific point in time. God has also predetermined before the foundation of the world that this Therapist, that is, you, treating the sick, are God's vessel for pouring out His Divine healing. You must believe that you, not somebody else; in this very moment, you are serving as an instrument for a Supernatural activity from God the Father Almighty through Jesus Christ by the power of The Holy Spirit. Look for an opportunity to share the Word. Talk about Jesus, about His saving grace. For He is the One who heals. Rebuke sin. Speak about how sin brought into the world sickness, pain, disease and death. Talk about how Jesus will put an end to sickness, disease, pain and death; there will be no more weeping, no more sorrow, no more pain. The healing we experience is merely a shadow of healing which lasts forever that Jesus gives to those who believe in Him. Jesus Himself said this, "For God loved the world so much that He gave His only Son, so that everyone who believes in Him will not die but have eternal life. For God did not send His Son into the world to be its judge, but to be its Savior. Whoever believes in the Son is not judged; but whoever does not believe has already been judged because he has not believed in God's only Son." (John 3:16-18)

Remember to put into practice the warning our Lord gave His disciples when He sent them on a mission to preach the gospel and heal the sick. Know that you and I are included. He says, "Listen! I am sending you out just like sheep to a pack of wolves. You must be as cautious as snakes and as gentle as doves. Watch out, for there will be men who will arrest you and take you to court..." (Matthew 10:16-17)

Do not share the word in a hasty, judgmental, overbearing, and self-righteous manner. It took God Himself six days to create the world and populate it. We are all sinners. As the Scriptures say, "There is no one who is righteous..." (Roman 3:10) Ask questions about the origin of sin. Sin came from one man. Refer to Romans 5:12, "Sin came into the world through one man, and sin brought death with it. As a result, death has spread to the whole human race because everyone has sinned." Death, as the bible says, is the price for sin. "For the wages of sin is death, but the gift of GOD is eternal life through Christ through Jesus Christ our Lord." (Romans 6:23) "But God has shown us how much He loves us – it was while we were still sinners that Christ died for us! By His sacrificial death we are now put right with God; how much more, then, will we be saved by Him from God's anger!" (Romans 5:8-9) How are we saved? That is, how do we obtain healing of our Spirits? "For whosoever shall call upon the name of the Lord shall be saved." (Romans 10:13)

"...Because, if you confess with your lips that

Jesus is Lord and believe in your heart that God raised Him from the dead, you will be saved. For man believes with his heart and confesses with his lips and so is saved. The scripture says, "No one who believes in him will be put to shame." (Romans 10:9-11) Your clients will consist of two types of people; the Just and the Unjust. Here is what the bible says in Habakkuk 2:4,"Behold the proud, his soul is not upright in him; but the Just shall live by his faith." (NKJ) The Just may be already brothers or sisters in Christ, or unbelievers who will be experiencing the touch of the Holy Spirit during the course of treatment. The unjust are those who may show a hostile attitude to anyone telling them about Jesus, or they may ridicule you and ask to see signs or proof. Their type was also around during the earthly Ministry of our Lord Jesus. This is what the bible says about them, "Then some teachers of the Law and some Pharisees spoke up. "Teacher," they said, "we want to see you perform a miracle." "How evil and godless are the people of this day!" Jesus exclaimed. "You ask me for a miracle? No! The only miracle you will be given is the miracle of the prophet Jonah. In the same way that Jonah spent three days and nights in the big fish, so will the Son of Man spend three days and nights in the depths of the earth." (Matthew 12:38-40)

The Holy Spirit is not like a faucet that can be turned on and off. The Holy Spirit is not for sale. The Holy Spirit is the Third Person of the Godhead. For He said to Moses, "I will have mercy on whom I have mercy, and I will have compassion on whom I have compassion." (Romans 9:15, and Exodus 33:19)

The client/sick/recipient must trust God for His gift of health, therefore must be ready to receive the gift of healing by faith. "Whoever doubts is like a wave in the sea that is driven and blown about by the wind. A person like that, unable to make up his mind and undecided in all he does, must not think that he will receive anything from the Lord." (James 1:6-8)

*****

I will take sickness away from among you
"You shall serve the Lord your God, and he will bless your bread and your water, and I will take sickness away from among you." Exodus 23:25

*****

17
Application

"If you heal me O Lord, I will be healed; if you save me, and I shall be saved; for thou art my praise." (Jer.17:14 KJV)

"I am the LORD, who heals you."(Ex.15:26)

FIRST BIBLICALLY RECORDED MASSAGE THERAPY

In Esther 2:9 and12 TEV, it is stated that the young Jewish orphan-cousin of Mordecai that would be future Queen of Persia, as part of her preparations for Persian empire-wide Beauty Pageant for the King - Ahasuerus to pick his wife - received several treatments of massage and special diet... the regular beauty treatment for the women lasted a year...massages with oil of myrrh for six month and with oil of balsam for six more..

Biblical Massage should be performed with the Client fully clothed without oil. But oil may be used on the upper and lower extremities and plantar, or when the Client requests draping, and the Minister/Practitioner determines it is necessary; then it must be performed with the help of an assistant or a deacon or deaconess present. Western Massage techniques, as well as those for special population, including, Sports Massage/Chair massage may be integrated with Proprioceptive Neuromascular Facilitation (PNF) , myofascial, lymphatic drainage, deep transverse friction and isometrics may be used according to the determination of the Minister/Practitioner regarding its relevance to the treatment needs of the client.

Duration: May vary from 10 minutes in respect of chair massage to as long as 2 hours, depending on preference. Client may be fully clothed.

Ambiance: Divine Presence with Christian Worship Music.

Lighting: Celestial blue

Scientific Argument: B. F. Skinner and Albert Bandura's, "Environmental reinforcement, and operant conditioning."

Preparation Before beginning massage, Minister/Therapist silently steps into the presence of God in humility with prayerful heart of thanksgiving and praise, through our Lord Jesus Christ for all God's Graces. He/she should continue to thank God for the gift of salvation through Christ Jesus, and healing manifesting in the life of the client at the touch of the Minister/ Therapist's hands by the power of the Holy Spirit so that God's name will be glorified (Psalm 100).He must reverently pray this prayer from Acts 4:29-30 , "Almighty God

Creator of Heaven and Earth, may you allow your servant to speak your message with all boldness. Reach out your hand to heal and grant that wonders and miracles may be performed through the name of your Holy Servant Jesus Christ." Amen. The Practitioner/Minister must firmly believe in the power and authority our Lord Jesus Christ has given him/her after He Resurrected, "to preach the Gospel...perform miracles...drive out demons...lay hands on sick people, and these will get well." (Mark chapter 16 verse 17) Amen. For the scripture has also said that Almighty God has pre-determined the roles of the Minister / Therapist and the client before the foundation of the world. So ask God for the power of the Holy Spirit to heal the sick in order that God's plan and purpose will be accomplished. Amen!

Steps

Invocation, Dedication & Exhortation by practitioner.

Needed: Client's affirmation of faith in Christ Jesus and acceptance of the notion that Healing is a gift by Almighty God the client's Creator through Jesus Christ by the power of the Holy Spirit, which is implicit in Salvation. Songs of Praises/ Worship (Choice as led by the Holy Spirit) Prayer, Western/Swedish Massage. Begin and end with "Passive Touch" with both hands very lightly on the temporalis (Client Seated or supine). About 3 minutes. No sound except healing worship hymn. Next, move one palm and place gently on the frontalis and the other on the pectoralis major without any movement. Next repeat softly, almost in a whisper Philippians 4:6-7 and ask client to mutter those words repeatedly to himself or herself with eyes closed; a powerful technique is for client to visualize Jesus Christ laying on his healing hands on him or her. If client is lying prone, he or she must observe Philippians 4:8. The practitioner must then proceed with appropriate technique or modality.

Your Prayer of Faith

Our Father in Heaven. Father of our Lord Jesus

Christ. I thank you for this privilege that I have from you to come boldly to your Throne of Grace even though I am a sinner. Lord God I know that if I died right now a sinner as I am, I would end up in hell. But because of your love for me you gave your only Son Christ Jesus to die for me. Lord Jesus Christ, I am trusting you, by faith, as my personal Savior and my only hope for heaven. You were crucified. You died. You were buried. God the Father Almighty rose you from the dead. You ascended into heaven from where you will come again to judge the living and the dead.

Come into my heart Holy Spirit and make me the new creation in Christ Jesus. Thank you for dying on the cross for me Lord Jesus and for saving me from spending eternity in hell. In your name Lord Jesus I pray. Amen.

Western Massage with Worship Hymns,
Healing Scriptures and words of prayers.
Lymphatic Drainage (If Necessary)
Myofascial Release'
Thanksgiving prayer with client. (Ending)
My Experiences.

I worked at an Accident Clinic in Tampa, Florida, over a period of ten months where I conducted a personal study to see how application of the effect of Biblical Massage affected the clients/ patients receiving that were referred to me by their physicians and chiropractors for therapy following an accident. I worked four days a week. On every given day, I treated ten patients a day. Each session lasted forty to sixty minutes. It was during the preliminary consultation period that I had an opportunity to gather important information concerning the client's reason for seeking Biblical Massage therapy, and, after reviewing information from the client's intake form about his or her preferences, or needs, I explained policies and "Biblical Massage and Holy Spirit Touch" procedures. This involved a full disclosure of my Christian belief, and explanation that my practice is grounded in the notion that "a person is healed only when he becomes one with God, and that this method of healing unites the physical, emotional, and the spiritual aspects of the person toward a more complete treatment," quoting Dr. John Chirban, Ph. D, Th. D, Director of the Institute of Medicine, Psychology, and Religion in Cambridge, Massachusetts, and clinical instructor in psychology at Harvard Medical School and The Cambridge Hospital, teaching courses in spirituality. What this means is that Biblical Massage and Holy Spirit Touch is a lifestyle, which, for some, begins with the therapy, and for others, helps to rekindle the client's indwelling Holy Spirit fire that gives life and health and peace. Biblical Massage and Holy Spirit Touch is an intensely Spiritual "treatment" technique that involves trusting Almighty God with healing the sick person's spirit mind, and body. The practitioner uses palpation and manipulation of soft tissue in Christ-centered ambiance while soft Christian hymns of praises, worship and healing music of worship hymns or gospel is played.

For those looking for evidence-based massage therapy, I will argue there is scientific explanation for it. It is called, "classical conditioning." Yes, it is explained by Pavlovian theory of classical conditioning, a phenomenon which is evident among cancer patients; the scientific world names this phenomenon, anticipatory nausea (AN). Tomoyasu, N, Bovbjerg, DH & Jacobson, PB (Feb. 1996), in a peer-reviewed article stated, "Anticipatory nausea in patients receiving emetogenic chemotherapy has been cited as an example of the importance of classical conditioning in clinical medicine" (Physiology Behavior). Evidence exist that cancer patients receiving cytotoxic drugs are subject to nauseous side effects. Studies showed that for outpatients, the thought of a clinic appointment is sufficient for the patient to start vomiting. This is classical conditioning. Ivan Pavlov, the original theorist had been feeding his dog immediately after he rang a bell; he later, found following a series of experiments that whenever he rang a bell , the dog would start salivating on hearing the sound of the bell even when food was not present. Pavlov termed this phenomenon as classical conditioning: it is the process of pairing a neutral stimulus such as the sound of a bell Pavlov called conditioned response (CR) with, for example, food; Pavlov named this an unconditioned stimulus, (CS) and named the taste of food conditioned stimulus (CS). He also named the involuntary physiological activity triggering salivating as unconditioned stimulus, and the response to the food as unconditioned response (UR). Salivating was involuntary for the dog. Pavlov's dog did not need to be trained to salivate; Pavlov found that the dog's response to the sound of the bell was automatic despite the absence of food. Classical conditioning is more than a dog's involuntary response to the tintinnabulation of the bell. It affects human behavior. find following tests of 55 cancer outpatients that, "Stimuli paired with the Unconditioned Stimuli (US) can provoke conditioned stimuli (CSs) eliciting anticipatory nausea (AN) as the conditioned response (CR)." (National Library of Medicine, National Institutes of Health.) Similarly, I would argue that Biblical Massage and Holy Spirit Touch elicits positive responses; while the conditioned response (CR) of cancer outpatients preparing for clinic appointment elicits anticipatory nausea (AN), with Biblical Massage and Holy Spirit Touch therapy for a client suffering emotional or muscular pain, the conditioned response (CR) elicits AH or anticipatory healing or health. The sanctuary-like ambience and the mood worship music are like the sound of the bell in Pavlov's experiment. Palpation and manipulation of client's soft tissue by practitioner while client meditates on things that are good and that deserve praise, things that are true, noble, right, pure, lovely, compare to conditioned stimuli (CS).

In my clinical experience while I was working for Healing Hands Medical Center, Tampa, Florida, and later while operating my own nonprofit Biblical Center, clients who were referred to me by their doctors had prescription to receive therapy at three sessions weekly for a three-month period. Before we started, I informed them that biblical massage and holy spirit touch was an experimental integrative holistic therapy aiming at healing the Spirit, mind, and body, and that the clients were free to opt out. Beginning each session with a client, I asked politely if it was all right that we prayed together. If they gave their consent as they always did, I turned on softly my favorite healing music from a CD by Benny Hinn; "Holy Spirit Thou Art Welcome In This Place," and I would then suggest to the client to lead in praying. If the client was reluctant to pray or replied that I proceeded, I complied. After the prayers, while I began "palpating and manipulating soft tissue," I turned on "I Am The One That Healeth Thee," also from Benny Hinn's CD and "Stand Up For Jesus" by Sarah Kelly from a CD entitled "Best Of Worship." My choices of worship and healing songs also included "Jesus Messiah," "Mighty To Save", "Merciful Savior", "Amazing Grace (My Chains Are Gone")", and "You Are My King Amazing Love" - all of which I obtained from a CD entitled "30 All Time Favorite Worship Songs". In the second week, I liked to start the seventh session a worship song by Geoff Bullock entitled "The Power of Your Love".

They helped to set the mood for the most reticent, hostile and uncooperative patients/clients. Some clients were so emotionally disturbed they treated a practitioner as if he or she was an enemy. I began the second week's sessions with the song Lord I Come to YOU. Here, in part, are the lyrics:

"Lord I come to You, Let my heart be changed,
renewed/ Flowing from the grace, That I've found in You/
 I have come to know weaknesses I see in me/
Will be stripped away/
By the power of Your love/
Hold me close/
Let Your love surrounds me /
Bring me near
Draw me to Your side ..."

I observed that by the sixth sessions of the second week, or the beginning of the third week, the clients started to participate along humming the praise or worship music while it was playing during massage.

However, there were other clients who reacted by venting out long-stored pent up emotional issues; this is a phenomenon known to massage therapists as "emotional release." Note that I had not yet attempted to share the Word because I knew that if the Holy Spirit was present, He would take control. So, in my silence I meditated continuously on the Word and quietly petitioned for His supernatural manifestation in the life of the client for the glory of God. It was in the third week while the CD was playing that a dramatic event was showing; clients who had previously indicated that they were either skeptics, agnostics or atheists were increasing getting more relaxed and feeling free to express their personal views on religion or ask questions relating to the bible, or Christianity, or about Jesus. That was the opening to lead them "to know Christ and the power of his resurrection and the fellowship of sharing in his sufferings, becoming like him in his death," as Apostle Paul wrote in his letter to Philippians 3:10. At the end of the three-month therapy, clients reported having experienced changes in their psychological and Spiritual dispositions as well. "Chuck," 26, (not his real name), had been in and out of prison 24 times on drug charges, became a believer. "Lawyer Allen", a self-professed "diehard skeptic", turned a follower of our Lord Jesus Christ and a bible student. "John", and his wife "Martina" , introduced John's mother who had been suffering a chronic lower back pain from a three-year old car accident said that he had trouble over carved objects his Brazilian mom carried along with her everywhere she went; she treated it with such special care that he suspected his mother was under demonic influences, and that he thought was the cause of her unremitting pain. Those were very few examples out of a total of forty clients in ten months. Before their prescribed three-month long three times a week therapy session were over, it became clear to me that I had become all things to my clients. So, I did everything I could to ensure that I did not conduct myself in any way beyond my scope of practice. I referred those in need of professional counseling with regard to their emotional problems to their appropriate professionals, and those who needed biblical guidance I directed to attend a church of their own choosing, but I also advised that they must check the theology and doctrine of that church to ensure it teaches biblical Christianity, and not prosperity gospel, motivational speech in the guise of Christianity or pastor worship.

Independent Supporting Scientific Studies

Although, there yet exists no independent scientific study to support the validity of Biblical Massage and Holy Spirit Touch as effective intervention for emotional or muscular pain, there are ample scientific evidences beside the Pavlovian classical conditioning argument to support the high probability of Biblical Massage Therapy and Holy Spirit Touch being evidence-based. The environment, methods and steps for Biblical Massage and Holy Spirit Touch I have specified combine methods in related studies on mental, cardiovascular, physical benefits of religion and spirituality from which researchers reported positive conclusions. Of relevance, is conclusion of a study in a peer reviewed article ISRN Psychiatry Journal published online, Koenig, Harold, G (2012), reported:

> "A large volume of research shows that people who are more R/S (Religious/Spiritual) have better mental health and adapt more quickly to health problems compared to those who are less R/S. These possible benefits to mental health and well-being have physiological consequences that impact physical health, affect the risk of disease, and influence response to treatment."

Koenig (2012) wrote that, "on January 3, 2009, after the death of the Guinness Book of World Records' oldest person, Maria de Jesus age 115, next in line was Gertrude Baines from Los Angeles. Born to slaves near Atlanta in 1894, she was described at 114 years old as "spry," "cheerful," and "talkative." When she was 112 years old, Ms. Baines was asked by a CNN correspondent to explain why she thought she had lived so long. Her reply: Institutes of Health) "God. Ask Him. I took good care of myself, the way he wanted me to." Brief and to the point." (ISRN Psychiatry, PMC, US National Library of Medicine National Institutes of Health).

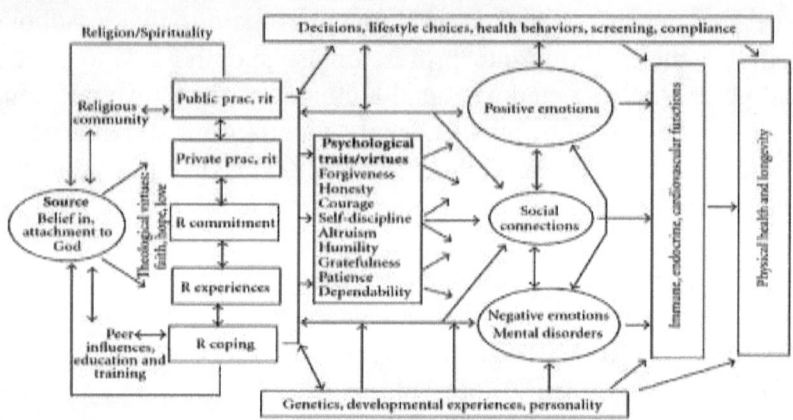

According to Koenig (2012), "Pain and Somatic Symptoms on the one hand, pain and other distressing somatic symptoms can motivate people to seek solace in religion through activities such as prayer or Scripture study. Thus, R/S is often turned to in order to cope with such symptoms. For example, in an early study of 382 adults with musculoskeletal complains, R/S coping was the most common strategy for dealing with pain and was considered the second most helpful in a long list of coping behaviors [531]. More recent research supports this earlier report [532]. On the other hand, R/S may somehow cause an increase in pain and somatic symptoms, perhaps by increasing concentration on negative symptoms or through the physical manifestations of hysteria, as claimed by Charcot in his copious writings around the turn of the 20th century [533].We identified 56 studies that examined relationships between R/S and pain. Of those, 22 (39%) reported inverse relationships between R/S and pain or found benefits from an R/S intervention, whereas 14 (25%) indicated a positive relationship between R/S and greater pain levels (13 of 14 being cross-sectional). Of the 18 best studies, nine (50%) reported inverse relationships (less pain among the more R/S [534] or reduced pain in response to a R/S intervention [535–542]), while three (20%) reported positive relationships (worse pain in the more R/S) [543–545]. Research suggests that meditation is particularly effective in reducing pain, although the effects are magnified when a religious word is used to focus attention [546,547]. No clinical trials, to my knowledge, have shown that meditation or other R/S interventions increase pain or somatic symptoms. The instructions provided in this book on how to perform Biblical Massage and Holy Spirit can form the framework for clinical trials for somatic symptoms because the technique integrates touch, meditation, prayer, music, belief.

"…religious beliefs have the potential to influence the cognitive appraisal of negative life events in a way that makes them less distressing. For people with medical illness, these beliefs are particularly useful because they are not lost or impaired with physical disability—unlike many other coping resources that are dependent on health (hobbies, relationships, and jobs/finances." (Koenig,2012)

18

HOW TO PERFORM BIBLICAL MEDITATION

Harvard Medical School Professor Emeritus, cardiologist, Herbert Benson, M.D, following ten years' researching the health benefits of transcendental meditation, (TM) , established the efficacy of meditation for healthcare, and named a meditation technique he developed, "The Relaxation Response," also named belief based meditation, "The Faith Factor." (The Relaxation Response, p.21) Benson (2000) wrote, "In my practice, I found that belief – and for many, this might mean religious belief – could not be divorced from the medical experience, as traditional medicine required. Belief was central to patients' lives, and potentially central to their health. Eighty percent of my patients chose prayers as the focus of their elicitation of the Relaxation Response. For this reason, I found myself in a curious position – that of a physician teaching patient to pray. By no means had I set out to do this. Patients' religious affiliations were as diverse as their ages and medical conditions, but they demonstrated to me the role that religious belief could play in healing. Remembered wellness (the placebo effect) appeared to enhance the effectiveness of the Relaxation Response. I called the combined force of these two internal influences "the faith factor." (The Relaxation Response, p.21)

I stumbled upon Biblical Meditation accidentally. I moved to Tampa, Florida, in 2004, from New York, because I had been so ill my body was not responding to treatment, and medical reasoning concluded I was on the brink of death. Please turn to pages 93 and 94 and read the account of Dr Jon E Hemstreet , my personal physician at Tampa General Hospital my health condition then. As an ordained minister, I found that reading the scriptures, focusing on the passages analytically in efforts to understand its messages often removed my thoughts about the ailment and numbed the pain. During one of those searches, reading Ezekiel chapter 47, a verse jumped up at me as if it was speaking directly to me. "On the banks , on both sides of the river, there will grow all kinds of trees for food. Their leaves will not wither nor their fruits fail, but they will bear fresh fruit every month, because the water for them flows from the sanctuary. Their fruit will be for food, and their leaves for healing," (Ezek.47:12). While I prayed for illumination and continued searching the scriptures, I kept noticing other passages and verses as though they were messages with my name on them. One was from Joshua chapter 1:8-9. It said the LORD appeared to Joshua and told him, "This book of the Law shall not depart out of your mouth; you shall meditate on it day and night, so that you may be careful to act in accordance with all that is written it . For then you shall make your way prosperous, and then you shall be successful. I hereby command you: Be strong and courageous; do not be frightened or dismayed, for the LORD your GOD is with you wherever you go." I believed I was Joshua. Every word of the text was for me. My doctors were not expecting me to live because the multi organ illness afflicting me refused the medications. I got to the point I trusted that if God's word urged me to strong and courageous, and not to be frightened or dismayed, I would trust and obey. My searches resulted in my discovering all together 120 plants mentioned in the bible. By the time that I went through Philippians chapter 4, I had an amazing revelation. Apostle Paul was revealing a mystery now being proven as scientific truth by the healthcare and medical communities; drugs don't cure. Compliance with biblical injunctions elicits activity of the parasympathetic nervous response to trigger the body's natural healing mechanisms. So, Apostle Paul writes from Roman prison, even though under intolerable stress, he urged, "Do not worry about anything ; but in everything by prayer and supplication with thanksgiving let your requests be made known to God. And the peace of God which surpasses all understanding will guard your hearts and your minds in Christ Jesus." (Phil.4:6-7) Now, Paul drops a bomb! Here is the key to the secret of Biblical Meditation, Paul appeared to say: "Whatever is true, whatever is honorable, whatever is just, whatever is pure, whatever is pleasing, whatever is commendable, if there is any excellence and if there is anything worthy of praise, meditate on these things. Keep on doing the things that you have learned and seen in me, and the GOD of peace will be with you. (Phil4:8-9) The things believers are urged to meditate on also include love, peace, patience, kindness , generosity, faithfulness, and self-control" (Gal.5:22).

RECAP

1. We perform Biblical Meditation because God commanded us to do so.
2. God gave his unconditional promise that when we keep meditating on his words day and night, we will make our way prosperous, and be successful.
3. Paul gave us God's rule of conduct: Rejoice in the Lord always and to let everyone see gentleness in our behavior.
4. Stop worrying. In other words, be anxious for nothing! Instead, ask God for our needs and giving him thanks for everything your heart will fill with the peace of God which is beyond everything we can ever think of.

This is how to do biblical meditation; it is different from the world's transcendental meditation which calls for emptying the mind. To do biblical meditation, we must fill our minds with whatever is true, whatever is honorable, whatever is just, whatever is pure, whatever is pleasing, whatever is commendable, if there is any excellence and if there is anything worthy of praise, meditate on these things. Apostle Paul says , "Keep on doing the things that you have learned and seen in me, and the GOD of peace will be with you." Those things include love, peace, patience, kindness, generosity, faithfulness, and self-control.

19
STEPS TO DO BIBLICAL MEDITATION

1. Begin by setting aside to pray and meditate. Give yourself at least one day of the week to reading a chapter from the Old Testament and a chapter from the New Testament. For, knowledge shows you how to meditate, what to meditate upon, and why meditate. Knowledge of the scripture teaches you how to pray, why pray, what to say in your prayers.
2. Duration: Any time from 15 minutes to 2 hours depending on individual's choice.
3. Ambiance: Divine Presence with Christian Worship. Music.
4. Lighting: Celestial blue
5. Scientific Argument: Environment influences behavior as in operant conditioning, according to B. F. Skinner and Albert Bandura.
6. Process: Play worship hymn or music in an ambience bathed in mood lights. I like blue soft light because the color blue lends to calmness when it is enhanced by the sound of soft worship music.
7. Start praying by giving thanks.
8. Bring your personal requests to him because the Lord Jesus said, Ask, and will be given.
9. Concentrate on the words of the worship hymn, or just the sound of the music, or take a scripture and focus on it.
10. Do these with eyes closed.
11. Do these while your breath rhythmically through the nose and exhale through your mouth.

MY OWN EXPERIENCES
Having been performing biblical meditation for the past 15 years from age 65, I am 80 now, and I no longer impose restriction on myself about how long or short duration to meditate, or about what to say in prayers. For there are times my meditation lasts for ten minutes, and other times two hours. But every day, before I put my hand on the doorknob to step outside, I pray Psalm 19:14 "Let the words of my mouth and the meditation of my heart, be acceptable in your sight, O God, my Redeemer and my Rock." In the evening , when I am about to go bed , I reflect on the words I spoke during the course of the day, whether I said or did things I ought not to have done, and left undone those things

that I ought to do. Then, I ask God's forgiveness through the Lord Jesus. I do not claim I have everything pat. I am 80 years of age now. It is February 2019. Last year on July 2, I suffered TIA or mini stroke. While I was on admission in Brandon Regional Hospital, I was diagnosed with intermittent arterial claudication, which means the main arteries in both of my legs are partially blocked , and doctors said I had 50/50 chances of recovery whether or not I chose to go into surgery. I said no to surgery. I was also diagnosed with carotid stenosis, meaning there is plaque blockage in an artery that carries blood into my brain. I was also diagnosed with thoracic aortic aneurysm, that means that my chest artery which takes blood from the heart into my body has ballooned can burst at any time. I have tried to live right, eat right, and love right, but I found that God's ways are not our ways, neither are our ways God's ways. I am disclosing this very personal information because somebody may learn from it and be encouraged when going through similar trials. At 80 years of age, I am getting ready to return to college for my master's degree in clinical psychology, and to join with a team in conducting research on some biblical concepts. I am excited!

A NOTE OF CAUTION
The purpose of this book is to inform and not to prescribe. Therefore, it is important that you consult your doctor, however you may feel. Biblical Massage and Holy Spirit Touch is contraindication, particularly, for vascular diseases; it is not contraindication for muscular and emotional pain. I am thankfully given this gem by my own trusted personal care physician of 15 years since 2008, Dr Orlando Rangel at 4160 North Armenia Avenue, Tampa, Florida, and his team of physicians Dr Andre Cintra, and Nurse Practitioner, Luz Marina Arias-Irrzarry in consultation with noted cardiologist, Dr Nieto, at Heart and Vascular Associates, 2727 MLK Blvd, and Vascular surgeon, Dr Venkataramanan Gangadharan , at Cardiology Center of Tampa,13701 Bruce B Downs Blvd, Suite 101, Tampa. At this point, I will quote myself as an advice from my first book I wrote following an illness a 15 years ago causing conventional medical opinion to conclude I was dying. In the book, God's Amazing Bible Plants Healed Me, I cautioned: "When we are seeking healing, we must flush from our system and our minds anything or anyone whose presence will cause us to waver in our faith in the LORD, or impede our relationship with Almighty God and His Son our Savior and Lord Jesus. Some of

such negative advices forbid any other form of healing except healing by faith. Others discourage the use of conventional medicine, while crediting herbal medicine alone as being the panacea for all ills. Do not do such a thing as follow such fallacies. Avoid such false teachings. Whether we are healed through conventional medical procedures, herbal healing, spiritual healing, they are all God-given. Some may work better than others because Almighty God ordained them that way. Sometimes we may find that a combination of all healing methods: conventional, herbal, dietary, physical exercises, fasting and prayers are the necessary ingredients for restoring our health. Never forget that Almighty God who bore all our diseases, and by whose stripes we are healed, the God who healed the sick by his word and by the power of his touch, the same God "is able to do exceeding abundantly above all that we ask or think, according to the power that works in us. "(Eph 3:20)

Recall wise counseling from the book of Sirach chapter 38 verses 1-15 "Give doctors the honor they deserve, for the Lord gave them their work to do. 2 Their skill came from the Most High, and kings reward them for it. 3 Their knowledge gives them a position of importance, and powerful people hold them in high regard.

4 The Lord created medicines from the earth, and a sensible person will not hesitate to use them. 5 Didn't a tree once make bitter water fit to drink, so that the Lord's power might be known? 6 He gave medical knowledge to human beings, so that we would praise him for the miracles he performs. 7-8 The druggist mixes these medicines, and the doctor will use them to cure diseases and ease pain. There is no end to the activities of the Lord, who gives health to the people of the world.

9 My child, when you get sick, don't ignore it. Pray to the Lord, and he will make you well. 10 Confess all your sins and determine that in the future you will live a righteous life. 11 Offer incense and a grain offering, as fine as you can afford. 12 Then call the doctor—for the Lord created him—and keep him at your side; you need him. 13 There are times when you have to depend on his skill. 14 The doctor's prayer is that the Lord will make him able to ease his patients' pain and make them well again. 15 As for the person who sins against his Creator, he deserves to be sick.

20
Christians, Reiki, Taoists and Qigong

How should a Biblical Massage Minister/Therapist/Practitioner relate with Qigong? From biblical perspective, a fruit does not fall far from a tree. Qigong is a product of Taoism as Christianity is to Judaism. Advocates for Taoism claim " Qigong is the predecessor of Christianity" because, Qigong, "the essence of Taoism as "Tao", is the "the way, the truth and the life", who is not the "eternal Tao", (not God the Father). Taoists say, "Tao" is not a person, it is a force, a principle." Qigong practitioners, like Taoists say, "The Tao regulates natural processes and nourishes balance in the Universe. It embodies the harmony of opposites (i.e. there would be no love without hate, no light without darkness, no male without female.)" But we, Christians point to the scriptures that say, "Jesus is the light of the world...The true light that enlightens every man...and, the world was made through him, yet the world knew him not." (John 8:12; John 1:9-10) Taoists believe, that life exists and is governed by cosmic duality of opposites formed by a circle divided into two equal sides of black and white with one containing a smaller circle of the opposite color: "Yin (dark side) and Yang (light side) representing the good and evil, light and dark, male and female. But we Christians have it on the authority of the bible that, "God is light, and there is no darkness at all in Him." (1 John 1:5) We know that darkness is produced by absence of light. Where there is light, darkness yields. Taoists do not have "a personified deity". But we know our God is anthropomorphic, and we have Jesus who came down to this earth and lived among men, and with whom we have a personal relationship. "Christ is the visible likeness of the invisible God. He is the first-born Son, superior to all created things. For, through him God created everything in heaven, and earth, the seen and unseen thing, including spiritual powers, lords, rulers, and authorities. God created the whole universe through him." (Colossians 1:15-16)

Taoists "do not pray as Christians do; there is no God to hear prayers or to whom to devote themselves." For us, Almighty God is our Father in Heaven. His Will is done on earth as it is done in Heaven. Taoists "seek answers to life's problems through inner meditation and outer observation." We, Christians "can do all things through Christ who strengthens" us. (Philippians 4:13 Taoists, like Qigong Masters say, "Yin and Yang are symbolic of cosmic dragon and tiger. They both religiously practice Tai Chi; the slow deliberate movements simulating the slithering sound and movement of a dragon, Taoists believe balance the flow of energy or "chi" within the body. Christians have no affinity with the dragon because the dragon is Satan. (Isaiah 27:01, Rev. 12:03 and Revelation 13:1-4) "Then I saw a beast coming up out of the sea. It had ten horns and seven heads; on each of its horns, here was a crown, and on each of its heads there was a name insulting to GOD. The beast looked like a leopard, with feet like a bear's feet and a mouth like a lion's mouth. The dragon gave the beast his own power, his throne, and his vast authority...Everyone worshiped the dragon because he had given authority to the beast. They worshiped the beast also..." (Revelation 13:01-04)

For I will restore health unto thee, and I will heal thee of thy wounds, saith the LORD; because they called thee an Outcast, [saying], This [is] Zion, whom no man seeketh after. (Jer 30:17)

Glossary

Active ingredients: The ingredients contained in any formula that give the desired physiological effect i.e. The components in a moisturizing cream that improves the moisture content of the skin.

Adrenal glands: Two organs situated one upon the upper end of each kidney. Stresses of modern life can exhaust the glands.

Acidophilus: A friendly bacteria found in the digestive system which combats the activities of invading micro-organisms associated with food poisoning and other infections.

AcidoAlphaHydroxyl Ceramides: Extract from sunflowers. A lipid that strengthens the skin's capacity to retain moisture, thereby supporting and sustaining skin's youthful smoothness and softness.

Acute: A short sharp crisis of rapid onset.

Adaptogen: A substance that helps the body to "adapt" to a new stress or strain by stimulating the body's own defensive mechanism.

Aetiology: A term denoting the cause or origin of a specific disease.

Agar-agar: Gelling agent made from seaweed.

Algae: A seaweed

Alginate: Gelatinous substance obtained from seaweed and used as an emulsifier and thickening agent.

Alkalis: Substances with a pH above 7. Often used as neutralizers in cosmetics and toiletries.

Alkaloids: Basic organic substances, usually vegetable in origin and having an alkaline reaction. Like alkalis they combine with acids to form salts. Some are toxic, insoluble in water.

Aloe Vera Extract: Effective healing agent and rich emollient. Used to counteract wrinkles, it is soothing and moisturizing.

Alteratives: Medicines that alter the process of nutrition, restoring in some unknown way the normal functions of an organ or system.

Allergy: Hypersensitivity to a foreign protein which produces a violent reaction eg. hayfever, asthma, irritable bowel.

Allopathy: Conventional medicine.

Amenorrhea: Suppression or normal menstrual flow during the time of life when it should occur.

Amino acids: Group of compounds containing both the carboxyl and the amino groups. They are the building blocks of proteins and are essential for the maintenance of the body.

Amphoteric: A normalizer "improve apparently contradictory symptoms".

Analgesics: Pain relievers,

Anodynes: Herbs taken orally for relief of mild pain

Anaphrodisiac: A herb that reduces excess sexual desire.

Antacids: Remedies that correct effects of stomach acid and relieve indigestion.

Antigens: substances, usually harmful, that when entering the body stimulate the immune system to produce antibodies.

Antibacterial: Any agent or process that inhibits the growth and reproduction of bacteria.

Antibody: A substance prepared in the body for the purpose of withstanding infection by viruses, bacteria and other organisms.

Anti-catarrhals: Agents that reduce the production of mucus.

Antifungals: Herbs that destroy fungi as in the treatment of thrush, candida etc.

Antihistamines: Agents that arrest production of histamine and which are useful in allergic conditions.

Annatto: A natural colorant derived from the seeds of a tropical tree.

Anthelmintics: Anti parasite.

Anthoposophical medicine: Holistic medicine based on the work of Dr Rudolf Steiner.

Antilithics: Agents used for elimination or dissolution of stone or gravel in bladder or kidney problems.

Anti-neoplastics: Herbs that prevent formation or destroy tumor cells .

Anti-pruritics: Agents to relieve intense itching.

Anti-spasmodics: Agents for relief of muscular cramps, spasm or mild pain.

Anti-tussives: Herbs that reduce cough severity, ease expectoration and clear the lungs.

Antioxidants: Substances that prevent the formation of free radicals which cause the oxidative deterioration that causes rancidity in oils or fats and also premature aging. Natural sources include vitamins A, C and E.

Aperient: Laxative

Aqueous coating: A natural water and vegetable cellulose coating which can be used as a coating to enhance tablet disintegration and dissolution.

Ascorbic acid: The chemical name for vitamin c.

Astringents: Products that cause a tightening and contractions of the skin tissues, generally used to tone skin and close pores. Can also arrest heavy bleeding.

Barrier cream: Cream that provides a protective coating when applied to the skin eg. hands and face.

Beeswax: A natural emulsifier and thickener.

Beta Carotene: An abundant source of Vitamin A with rich anti-oxidant properties. It is necessary for tissue repair and maintenance and accelerates the formation of healthy new skin cells. Vitamin A deters excess dryness.

Bisoprolol: Main active ingredient in chamomile which has excellent skin healing properties.

Bitters: Stimulate the autonomic nervous system. Bitters increase appetite, assist assimilation.

Botanical Extract: An extract of herbs and plants. The extracting solvent can be water, oil, alcohol or any synthetic solvent such as propylene glycol.

Broncho-dilators: herbs that expand the clear space within the bronchial tubes, opening up airways and relieving obstruction.

Candida albicans: A yeast that causes thrush and in more severe cases, symptoms can affect the whole body.

Capricin: A caprylic acid formulation that facilitates absorption of calcium and magnesium.

Caramel: Coloring agent derived from liquid corn syrup.

Carcinogens: Substances that bring about a malignant change in body cells.

Carmine: Natural red pigment obtained from cochineal.

Carminatives: Anti-flatulent, aromatic herbs used to expel gas from the stomach and intestines.

Catabolism: An aspect of metabolism which is concerned with the breaking down of complex substances to simpler ones, with the release of energy.

Cetyl Alcohol: Derived from coconut and palm oils. This is not a drying alcohol. Used as an emollient and to protect skin from moisture loss.

Chlorophyll: stored energy of the sun. Green coloring matter of plants.

Cholagogues: A group of agents which increases the secretion of bile and its expulsion from the gall bladder.

Choleretic: An agent which reduces cholesterol levels by excreting cholesterol.

Citric Acid: Derived from citrus fruit. A natural preservative that helps to adjust the pH of cosmetic products.

Clay: Deep-cleansing and highly absorbent. Bentonite and green clay are two types of natural clay.

Compresses: External applications to soften tissue, allay inflammation or alleviate pain.

Contra-indicated: Not indicated, against medical advice, unsuitable for use.

Cornstarch: Used as a safe base for our eye shadows, blushers and loose powders.

Coumarins: Powerful anti-coagulant plant chemicals. Used to prevent blood clotting.

Counter-irritant: An agent which produces vaso-dilation of peripheral blood vessels by stimulating nerve-endings of the skin to generate irritation intended to relieve deep-seated pain.

Cramp: Sustained contraction of a muscle.

Decoction: A preparation obtained by bringing to the boil and simmering dense herbal materials i.e. bark, root and woody parts for a plant to extract active constituents.

Decongestant: Herb which is used to loosen mucus within bronchi and lungs.

Demulcent: Anti-irritant. A herb rich in mucilage that is soothing, bland, offering protection to inflamed or irritable mucous surfaces.

Depurative: Blood purifier. Alterative.

Detoxifiers: Plant medicines that aid removal of a poison or poisonous effect, reducing toxic properties.

Diaphoretics: Herbs that induce increased perspiration.

Diuretics: Agents that increase the flow of urine from the kidneys and so excrete excess fluid from the body.

Douche: A term used to describe the cleansing of certain parts of the body eg. Washing wounds and ulcers, eye douches etc.

Eliminative: A herb to disperse and promote excretion from the body.

Emetic: A herb to induce vomiting. Given to expel poisons.

Emmenagogues: Plant substances which initiate and promote the menstrual flow.

Emollient: Any substance that prevents water loss from the skin. Most natural oils. perform this function.

Emulsifier: A substance that holds oil in water or water in oil. They are necessary in the manufacture of cream and lotions.

Emulsion: A mixture of two incompatible substances. Most creams on the cosmetic market are emulsions.

Enzyme: A biological catalyst that acts to speed up chemical reactions. **Digestive enzymes** are necessary for the breakdown of proteins, carbohydrates and fats i.e. pepsin.

Enuresis: Bed wetting.

Essential Fatty Acids: A fatty acid that must be supplied in the diet as the body cannot produce it itself.

Expectorants: Herbs that increase bronchial mucous secretion by promoting liquefaction of sticky mucous and its expulsion from the body.

Fatty Acid: A monobasic acid containing only the chemicals carbon, hydrogen and oxygen. Found in vegetable and animal fats, they are important for maintaining a healthy skin and are excellent emollients.

Febrifuge: Anti-fever

Fixed Oil: A fixed oil is chemically the same as a fat, but is generally liquid i.e. Almond oil, grapeseed oil.

Flavonoids: Natural chemicals that prevent the deposit of fatty material in blood vessels.

Flaxseed Oil: Rich source of omega 3 essential fatty acids. (Also known as Linseed Oil)

Fructose: A natural sugar found in honey and fruits.

Fumigant: Herb, usually a gum, which when burnt releases mixed gases into the atmosphere to cleanse against air borne infection eg myrrh or frankincense.

Galactagogue: Herb to increase flow of breast milk in nursing mothers.

Glycerin (vegetable): A humectant and emollient, it absorbs moisture from the air, thereby keeping moisture in your skin.

Glycoside: An organic substance which may be broken into two parts, one of which is always sugar.

Grain Alcohol: A natural solvent that evaporates easily.

Green Tea Extract: Works towards lightening the skin by actively slowing the transport of melanin to the skin's surface. Has well known antioxidant qualities.

Guar Gum: A fiber derived from the guar plant and used as a binder in tablet manufacturing.

Haemostatics: Agents that arrest bleeding.

Hepatic: A herb that assists the liver in its function and promotes the flow of bile.

Histamine: A chemical released via the body's immune system in response to allergens.

Hypoallergenic: In the strictest sense means without fragrance, but more broadly refers to products that are unlikely to cause skin irritation.

Hyaluronic Acid: Derived from yeast cells, it is extremely hydroscopic. Binds water in the interstitial spaces between skin cells, forming a gel-like substance which holds the cells together.

Hydrogenated Palm Kernel Oil: Source of essential fatty acids.

Infusion: The liquid resulting from making a herbal tea.

Iron Oxides: Natural mineral derived color pigments.

Kaolin: Clay used to absorb oils.

Laxative: Agent used for persistent constipation to help expel faecal matter from the bowel.

Lecithin: Natural antioxidant and emollient. High in essential fatty acids. A stabilizer and emulsifier from Soya beans, corn, peanuts and egg-yolk. Cholesterol reducer.

Lymph: A straw colored fluid which circulates many tissues of the body and serves to lubricate and cleanse them.

Magnesium Carbonate: Mineral derived from dolomite, it is used in our face powders to achieve the correct shade or tone.

Mannitol: A natural sugar substitute derived from the manna plant and seaweed.

Menorrhagia: Abnormally heavy menstrual bleeding, more than normal flow and longer lasting.

Menthol: Nature constituent of peppermint oil. Used for its antiseptic properties.

Metabolism: The reactions involved in the building up and decomposition of chemical substances in living organisms.

Metrorrhagia: Bleeding from the womb between periods.

Mineral Salts: Used for color pigments in our temporary hair color.

Mucilage: A slimy product formed by the addition of gum to water.

Used internally and externally to soothe irritated and inflamed membranes and surfaces.

Mucolytics: Agents that disperse or dissolve mucus.

Natural glycerine: Used to stabilize and disperse liquid
nutrients inside a capsule. A clear colorless syrupy liquid with a sweet taste derived from natural fats and oil.

Nerve restoratives: Herbs used to provide support and
restoration of the nervous system caused by stress, disease or faulty nutrition.

Nutrient: A non-irritating, easily digested agent which
provides body nourishment and stimulates metabolic processes.

Orexigenic: A herb which increases or stimulates the appetite.

Oxytocic: A herb which hastens the process of childbirth by initiating contraction of the uterine muscle.

Parabens (parahydroxybenzoic acid esters): A family of neutral, broad-spectrum antibacterials which have been used extensively for many years in the food, cosmetic and pharmaceutical industries as mild preservatives and have not been tested on animals for a long time. They are found in nature, but the ones used in cosmetics are synthetically produced. They have a long history of relatively safe use. Like all synthetic components they are used minimally and only when necessary. Effective levels are 0.1 - 0.3% concentration in the overall product.

Peripheral: Refers to the outermost parts of the body.

Poultice: Poultices are packs of powders, dried or fresh herbs, enclosed in a muslin bag or wrapped in folds of a flannel or linen and soaked in boiling water, then applied to the affected area of the body.

Prostaglandins: Hormone like messengers in the body responsible for the control of important body functions.

Proteinuria: Presence of albumin in the urine.

Pruritus: Itching.

Purgative: An agent that encourages evacuation of matter from the bowel.

Rice Bran Wax: Derived from rice bran, used as emollient and to protect skin from moisture loss.

Reflux: A backward flow of food to the mouth from the gullet or stomach.

Refrigerant: A cooling preparation taken orally or applied externally.

Resin: A thick-solid, insoluble in water but soluble in alcohol, exude from trees or plants which are used as antiseptics.

Rubefacient: External use. An agent to draw a rich blood supply to the skin, increasing heat to the tissues to aid the body in absorption of properties from creams, lotions etc.

Rose Essential Oil: Soothing, harmonizing effect on the skin. Considered the 'queen' of all essential oils, its gentle yet powerful nature has the ability to help repair broken capillaries.

Salicylates: Salts of salicylic acid sometimes used in rheumatism, gout and acid conditions.

Saponins: Constituents of some plants that produce soap like frothing effect when agitated in water.

Sedatives: Herbs that relax the central nervous system.

Seasalt Minerals: Thickener and disinfectant in shampoos.

Sesame Oil: Rich emollient properties and provides natural sun protection.

Sialagogue: Herbs that increase production of saliva and assist digestion of starches.

Silica (hydrated): A purified mineral, used as an anti-caking agent in the production of vitamin tablets.

Sodium Laureth Sulfate: Used as a naturally derived surfactant from coconut oil to make hair care and bath care products foam.

Sorbitol: Gives a velvety feel to the skin. Derived from cherries, plums, pears, apples and seaweed.

Spasmolytic: Another name for anti-spasmodic.

Stimulants: Herbs that spur the circulation, increase energy and physical function.

Styptic: A substance that stops bleeding usually by contracting the tissue.

Sudorifics: Similar to diaphoretics but are used to stimulate more profuse abundant sweating.

Systemic: Referring to the whole of the body.

Tannins: Present in tea and coffee and many herbs. Coagulate protein and inhibit the laying down of fatty deposits. Astringent.

Urinary antiseptic: A germicidal action of a herb destructive to harmful bacteria in the urine when excreted from the body via the kidneys, bladder and ureters.

Urinary demulcent: A soothing anti-irritant used for the protection of sensitive surfaces of the kidney tubules and ureters against friction, irritation.

Urinary haemostatics: Urinary astringents that arrest bleeding from the kidneys.

Vasoconstrictors: Agents that constrict blood vessels causing an increase in blood pressure.

Vasodilators: Herbs that promote dilation of the blood vessels causing a reduction of blood pressure.

Vegetable cellulose: Substance derived from various plant fibers, used as filler, and disintegrant in the production of tablets.

Vermifuge: A substance that expels or destroys intestinal worms.

Vesicant: A blistering agent.

Vitamin A: Potent anti-oxidant, used in combination with vitamins E and C as a natural preservative. Necessary for tissue repair and maintenance and accelerates the formation of healthy new skin cells. It benefits the treatment of skin disorders and oxidant, used as a natural preservative. Antiinflammatory properties aid in healing.

Vitamin E: A powerful natural anti-oxidant, used in combination with vitamins A and C as a natural preservative. It slows signs of aging and the degeneration of skin cells.

Vulnerary: A plant whose external application has a cleansing and healing effect on open wounds, cuts and ulcers by promoting cell repair.

Wheatgerm Oil: Rich in vitamin E, penetrates well to prevent loss of moisture and benefit cells.

Zinc Oxide: A chemical compound, ZnO, which is nearly insoluble in water but soluble in acids or alkalis. It occurs as white hexagonal crystals or a white powder commonly known as zinc white. Zinc white is used as a pigment in paints; less opaque than lithopone, it remains white when exposed to hydrogen sulfide or ultraviolet light. It is also used as filler for rubber goods and in coatings for paper. Because it absorbs ultraviolet light, zinc oxide can be used in ointments, creams, and lotions to protect against sunburn.

References

Blake and Edmonds, Biblical Sites in Turkey,

Carol Ann Rinzler. Nutrition For Dummies. Wiley Publishing

Constas, Nicholas. "An Apology for the Cult of Saints in Late Antiquity: Eustratius Presbyter of Constantinople, On the State of Souls After Death." Journal of Early Christian Studies 10 (2002): wikipedia dot com

Chirban, John T. Orthodox Theological Roots Of Holistic

Chirban, John T. "The Path of Growth and Development in Eastern Orthodoxy," in Sickness or Sin? Spiritual Discernment and Differential Diagnosis. Brookline, MA: Holy Cross Orthodox Press, 2001

Christopher Hobbs. Herbal Remedies For Dummies

Donald De Marco, Love and Healing world wide web the interim dot com/donald-demarco

Enns, Paul (1989), The Moody Bible Handbook

Rudwick and Green 1958 ;The Anchor Bible Dictionary

John McRay, Archaeology And The New Testament

Blueletterbible.org

Halley, henry H,(1925) Halley's Bible Handbook

Hezekiah K.Scipio (2007) : God's Amazing Bible Plants Healed

Sarah K. Yeomans (2011)Biblical Archaeologist Review and the TravelStudy

John Piper, Spiritual Gifts, Bethlehem Baptist Church sound of grace dot com

Louis, Gregory (2014), Reconstruction picture. Retrieved from https://wortsnall.files.wordpress.com/2014/09/roman-medicine-artists-reconstruction.jpg

Miller, Thomas S. The Birth of the Hospital in The Byzantine Empire, Baltimore, Md.: Johns Hopkins University Press,1985.

Koenig (1996), Religion, Spirituality, and Health : The Research and Clinical Implications. Departments of Medicine and Psychiatry, Duke University Medical Center, P.O. Box 3400, Durham, NC 27705, USA https://www.ncbi.nlm.nih.gov/pmc/articles/PMC3671693/

Nutton, Vivian. "From Galen to Alexander, Aspects of Medicine and Medical Practice in Late Antiquity," in Symposium on Byzantine Medicine, ed.

John Scarborough. Dumbarton Oaks Papers 38 (Washington, D.C.: Dumbarton Oaks Research Library and Collection, 1984)

Tai Chi utah dotedu/stc/tai-chi/articles dot html Taoism
taoist dot org
Merrill F. Unger . Unger's Bible Dictionary. Moody Press, Chicago, Ill. 1965
Ney-Grimm, J.M(n.d)   Roman meds discovered in 2nd century BC shipwreck
https://wortsnall.wordpress.com/2014/10/06/herbal-archaeology-roundup-2/

Perlman A, Fogerite SG, Glass O, et al. Efficacy and safety of massage for osteoarthritis of the knee: a randomized clinical trial. Journal of General Internal Medicine. December 12, 2018.

The Holy Bible, The Old and New Testaments with the Apocryphal/Deuterocanonical Books, New Revised Standard Version. American Bible Society, NY
The Holy Bible, Revised Standard Version. A.J.Holman Company, Philadelphia
The Holy Bible. King James Version. A.J. Hollman Company, Philadelphia
Henry H. Halley. Halley's Bible Handbook. Zondervan Publishing House, Grand Rapids, Michigan. 1962
The Good News Bible Today's English Version. American Bible Society, NY. 1985
Isadore Rosenfeld, M.D, Random House, NY 1996
Miriam Polunim. Healing Foods. DK Publishing Inc. 1997
Michael Castleman. The New Healing Herbs.
Isamu Sekido. Fruits, Roots And Fungi Plants We Eat. Lerner Publication Co. 1985
Russell Wild. The Complete Book Of Natural Medicinal And Cures. Rodale Press Inc., Emmaus, Penn.
Penelope Ody. The Complete Medicinal And Herbal Remedies
Lavon J. Dunne. Nutrition Almanac. McGraw-Hill Books
-Lesley Bremness. Herbs. Dorling Kindersley Books, NY.NY, 1994
-Gillian Roberts. Trees. Eyewitness Handbook, Allen J.Coombes, Dorling Kindersley, Inc, NY. 1992.
-Judith Sumner. The Natural History Of Medicinal Plants
-Anne McIntyre. The Medicinal Garden, How to Grow and Use Your Own Medicinal Herb, Henry Holt and Company, NY 1997, ISBN

0-8050-4838-3

Bentley, Robert and Henry Trimen. Medicinal Plants. London, Churchill, 1880. WZ 295 B556m 1880 Ken Fern

F. Chittendon. RHS Dictionary of Plants plus Supplement. 1956 Oxford University Press 1951

Hedrick. U. P. Sturtevant's Edible Plants of the World. Dover Publications 1972 ISBN 0-486-20459-6

Grieve. A Modern Herbal. Penguin 1984 ISBN 0-14-046-440-9.

Launert. E. Edible and Medicinal Plants. Hamlyn 1981 ISBN 0- 600-37216-2

Holtom. J. and Hylton. W. Complete Guide to Herbs. Rodale Press 1979 ISBN 0-87857-262-7

Simons. New Vegetable Growers Handbook. Penguin 1977 ISBN 0-14-046-050-0

Philbrick H. and Gregg R. B. Companion Plants. Watkins 1979

Riotte. L. Companion Planting for Successful Gardening. Garden Way, Vermont, USA. 1978 ISBN 0-88266-064-0

- Lust. J. The Herb Book. Bantam books 1983 ISBN 0-553-23827-2

Stockhorst, U, Klosterhalfen, S, Klosterhalfen, W, Winkelmann, W, and Steingrueber,M (1993), <u>Physiology Behavior.</u> 1996 Feb;59(2):273-6. <u>US National Library of Medicine</u> <u>National Institutes of Health</u> https://www.ncbi.nlm.nih.gov/pubmed/8838605

Thompson. B. The Gardener's Assistant. Blackie and Son. 1878

- Uphof. J. C. Th. Dictionary of Economic Plants. Weinheim 1959

- Hatfield. A. W. How to Enjoy your Weeds. Frederick Muller Ltd 1977 ISBN 0-584-10141-4

- Cooper. M. and Johnson. A. Poisonous Plants in Britain and their effects on Animals and Man. HMSO 1984 ISBN 0112425291

Mills. S. Y. The Dictionary of Modern Herbalism. 0

Facciola. S. Cornucopia - A Source Book of Edible Plants. Kampong Publications 1990 ISBN 0-9628087-0-9

Huxley. A. The New RHS Dictionary of Gardening. 1992. MacMillan Press 1992 ISBN 0-333-47494-5

-Allardice.P. A - Z of Companion Planting. Cassell Publishers Ltd. 1993 ISBN 0-304-34324-2

- Foster. S. & Duke. J. A. A Field Guide to Medicinal Plants. Eastern and Central N. America. Houghton Mifflin Co. 1990 ISBN 0395467225

- Thomas. G. S. Perennial Garden Plants J. M. Dent & Sons, London. 1990 ISBN 0 460 86048 8

Bown. D. Encyclopedia of Herbs and their Uses. Dorling Kindersley, London. 1995 ISBN 0-7513-020-31

Bentley, Robert and Henry Trimen. Medicinal Plants. London, Churchill, 1880. (WZ 295 B556m 1880)

Zondervan NIV Study Bible. Zondervan, Grand Rapids, Michigan. 1985. ISBN 0-310-92306-9

www.ingramcontent.com/pod-product-compliance
Lightning Source LLC
Chambersburg PA
CBHW030842180526
45163CB00004B/1421